RSC Paperbacks

RSC Paperbacks are a series of inexpensive texts suitable for teachers and students and give a clear, readable introduction to selected topics in chemistry. They should also appeal to the general chemist. For further information on all available titles contact:

Sales and Customer Care Department, Royal Society of Chemistry,
Thomas Graham House, Science Park, Milton Road, Cambridge CB4 0WF, UK
Telephone: +44 (0)1223 432360; Fax: +44 (0)1223 423429; E-mail: sales@rsc.org

Recent Titles Available

The Chemistry of Fragrances
compiled by David Pybus and Charles Sell
Polymers and the Environment
by Gerald Scott
Brewing
by Ian S. Hornsey
The Chemistry of Fireworks
by Michael S. Russell
Water (Second Edition): A Matrix of Life
by Felix Franks
The Science of Chocolate
by Stephen T. Beckett
The Science of Sugar Confectionery
by W. P. Edwards
Colour Chemistry
by R. M. Christie
Beer: Quality, Safety and Nutritional Aspects
by E. Denise Baxter and Paul S. Hughes

Future titles may be obtained immediately on publication by placing a standing order for RSC Paperbacks. Information on this is available from the address above.

RSC Paperbacks

BEER: QUALITY, SAFETY AND NUTRITIONAL ASPECTS

E. DENISE BAXTER

Brewing Research International,
Lyttel Hall, Nutfield, Redhill, Surrey RH1 4HY, UK

PAUL S. HUGHES

Heineken Technical Services,
Burgemeester Smeetsweg 1, 2382 PH Zoeterwoude,
The Netherlands

RS•C

ROYAL SOCIETY OF CHEMISTRY

ISBN 0-85404-588-0

A catalogue record for this book is available from the British Library

Published by The Royal Society of Chemistry,
Thomas Graham House, Science Park, Milton Road,
Cambridge CB4 0WF, UK
Registered Charity Number 207890

For further information see our web site at www.rsc.org

Typeset in Great Britain by Vision Typesetting, Manchester
Printed by Bookcraft Ltd, UK

Preface

Beer has been a popular beverage for thousands of years and brewing is often described as the oldest biotechnological process. Over the years the brewmaster's art has been supplemented by vast increases in our knowledge of the chemistry and biochemistry both of the ingredients and of the changes taking place to those ingredients during brewing. Together these contribute to give the products we recognise today – a wide range of different but consistently high quality beer types.

This book aims to explain the scientific principles which underpin those aspects of beer which are of the great interest to the beer drinker – namely its taste, appearance and nutritional qualities. This book is very much a synthesis of the current thinking as many aspects of beer quality are still tantalisingly elusive, so the story cannot be completed at the moment. . . .

Contents

Chapter 3
Flavour Determinants of Beer Quality   40

Glossary

α-Acids: The major constituent of the resin (humulones) in hop cones: α-acids are converted to bittering substances (iso-α-acids) during wort boiling.

Adjunct: Any source of fermentable extract other than malted barley used in the mash tun or the copper. May be solid, *e.g.* cereal grits, or liquid *e.g.* sugar syrup.

Air rest: An interruption of the steeping process to allow the barley to absorb oxygen from the air and thus to overcome water sensitivity and to ensure even germination.

Ale: Originally an unhopped but fermented malt drink, the term ale nowadays refers to any beer produced at temperatures of between 16 and 21 °C (most frequently around 18 °C) using a top-fermenting yeast (*Saccharomyces cerevisiae*).

Aleurone: The thick layer of living cells which surrounds the starchy endosperm in mature barley kernels.

Amylopectin: The second major constituent of barley starch, amylopectin is a large, highly branched molecule consisting of glucose units linked by α-1,4 and α-1,6 bonds.

Amylose: One of the two main components of barley starch. Amylose consists of a linear chain of glucose molecules linked by α-1,4 bonds.

Attentuation: The reduction in density of wort which occurs during fermentation as sugars are converted to alcohol.

Beer: In the UK, the legal definition of beer is for Excise purposes, and defines beer as any liquor made or sold as beer. The clearest technical definition describes beer as a fermented liquor produced mainly from malted barley but including other carbohydrate sources and flavoured with hops.

Cask: A large container for draught beer, originally made of wood, but now may also be made of aluminium. Traditionally, beer casks came in seven sizes: butt (108 gallons), puncheon (72 gallons), hogshead (54 gallons), barrel (36 gallons), kilderkin (18 gallons), firkin (9 gallons) and pin (4.5 gallons). *NB* 1 gallon = 4.54 litres.

Cold break: The precipitate formed when wort is cooled to room temperatures, consisting mainly of protein.

Copper: The vessel in which wort is boiled with hops to obtain the characteristic bitter flavours. So-called because it traditionally was made of copper, now often made of stainless steel. Also known as the kettle.

Crystal malt: Malt whose endosperm has been converted to a sugary crystalline mass during kilning. A proportion of crystal malt is added to the grist to provide colour and flavour to certain beers, particularly British ales.

Cylindroconical vessel: A cylindrical vertical tank with a conical base in which the yeast sediments after fermentation. Temperature is controlled by cooling-coils around the walls. Capacity ranges from 200 to 6000 hectolitres.

Embryo: The part of the barley kernel which gives rise to the new plant.

Endosperm: The part of the barley kernel other than the embryo. The endosperm consists essentially of a store of food for the new barley plant.

Finings: Charged colloidal substances, prepared from isinglass (collagen) from the swim bladders of certain tropical fish.

Flocculation: The clumping together of yeast cells at the end of fermentation. Also used to describe the clumping together of protein precipitated during wort boiling.

Germination: The sprouting of the resting barley seed to form new roots and shoots. The first visible sign is the cream-coloured 'chit' or first root emerging from the embryo end of the barley kernel.

Gibberellins: Natural plant hormones (phytohormones) produced by the barley embryo in response to steeping in water. Gibberellins stimulate the production of enzymes in the endosperm which hydrolyse the stored food reserves in the embryo and make them available to the growing plant.

Green beer: Freshly produced beer immediately after the end of primary fermentation and before conditioning (maturation).

Green malt: Barley germinated for between one and five days, before kilning, with a moisture content of at least 40%.

Grist: The term given to the mixture of coarsely ground malted barley, together with milled raw cereals and speciality malts (and barley) such as crystal malt or roast barley. Includes liquid adjuncts such as syrups. May also be applied to the mixture of hops and hop pellets added to the copper.

Hops: A perennial climbing vine, *Humulus lupulus*, a member of the family of *Cannabinaceae*. First recorded use to flavour beer was in Egypt, 600 years BC. The part traditionally used in brewing is the hop cone, which is the female ripened flower. In modern brewing, the hop cones are either extracted or finely powdered and compressed to form hop pellets which keep better and are easier to transport.

Hordein: The main component of barley protein. Closely related to similar proteins in wheat (gliadins), rye (secalins) and maize (zeins).

Hot break: Term given to the precipitate of protein which forms in boiled wort when it is cooled. Also called trub.

Husk: The outer, protective layers of the barley kernel, formed from the fruit and seed coats.

Isinglass: Collagen from the swim bladders of certain tropical fish, used as finings (*qv*) in beer to assist clarification.

Kettle: Another term, originally American, for the vessel in which wort is boiled. See also 'copper'.

Kilning: The final stage of malting in which the green malt is dried and cured by heating in a draught of warm air. The final temperature depends upon the type of malt being made.

Lager: A pale straw coloured beer produced from a lightly kilned malt and fermented by bottom-fermenting yeast (*Saccharomyces carlsbergensis*) at a low temperature (7–13 °C) and matured for several weeks.

Lautering: The process by which the sweet wort is separated from the spent grains, by drawing it off through the bed of spent grains.

Lauter tun: Vessel in which wort is separated from the spent grains by filtration through the spent grain bed. Generally a wide shallow vessel fitted with rakes to break up the bed.

Mashing: Process in which milled malt is mixed with hot water to extract cereal components, mainly starch. This starch is then converted to fermentable sugars by enzyme action.

Mash tun: The vessel in which mashing occurs. May also be called the 'conversion' vessel. In traditional ale brewing, the wort is also separated from the spent grains in the mash tun. However, in modern practice, it is more common to transfer the mash to a specific filtration vessel, the lauter tun (*qv*).

Original gravity (OG): This is the gravity of the wort prior to fermentation. In general, the higher the gravity, the more alcohol is produced, but there is no absolute correlation since worts may contain varying proportions of unfermentable material (such as protein). In addition, some types of beers retain some sugars that are potentially fermentable. The OG has often been the basis for calculating the excise duty payable, but nowadays the final alcohol content is more generally used.

Paraflow: A plate heat exchanger for cooling wort after boiling. Also used to cool beer before packaging.

Primings: Sugar added after the primary fermenatation, particularly to traditional mild ales and sweet stouts, to add some sweetness. May also be added to cask ales to provide additional fermentable extract for secondary fermentation in the cask.

Racking: The process of filling beer into casks, kegs or storage tanks after fermentation.

Small beer: A light, digestible table beer, relatively low in alcohol (OG < 1025°) produced from the Middle Ages by re-extracting grist already partially extracted to produce a strong ale.

Sorghum: A small-grained cereal grown in Africa and southern USA which can be used for brewing beer.

Spent grains: The residue of milled malt left after mashing. Spent grains consist mainly of husk and bran layers. They are relatively rich in protein and are used as cattle feed.

Steeping: The first stage of the malting process. Involves soaking the barley grain in water until the moisture is raised from 12% to 45%. Generally involves two or more immersion stages separated by air rests (*qv*).

Stillage: A wooden or metal structure which supports beer casks in a horizontal position in the cellar prior to dispense, allowing the yeast and protein to sediment with the finings and clarify the beer.

Trub: The coagulated protein which separates out in the wort after boiling. Also known as the 'hot break', the word trub is derived from a German word meaning 'break'.

Tun: A term used to describe any large vessel in a brewery, *e.g.* mash tun, lauter tun *etc*.

Whirlpool: A type of centrifuge used to separate the hot break or trub from the wort on cooling.

Wort: The sweet syrupy liquid which results from extraction and hydrolysis of starch from malted barley during mashing. After the addition of hops during boiling, sweet wort becomes bitter wort.

Yeast: A single-celled microorganism which, in the absence of oxygen, can use glucose as a respiratory substrate and convert it to ethanol. The two main strains used in brewing are *Saccharomyces carlsbergensis* (bottom fermenting lager yeast) and top fermenting ale yeast *Saccharomyces cerevisiae*. Individual brewing companies each have their own sub-strain, selected over countless generations for particular properties regarded as desirable to the brewer.

An Overview of the Malting and Brewing Processes

MALTING

The story of beer starts with ripe barley grain, plump and sound, with a moderate (for a cereal) protein content of 10–12%. The barley kernel is roughly ovoid in shape, surrounded by protective layers of husk, with a small embryo at one end. This embryo is the part that will grow into the new plant, given the chance. The remaining part of the kernel is the endosperm, which is basically just a store of food for the young plant.

Most of the endosperm consists of large dead cells with thick cell walls consisting mainly of β-glucan (a polymer of glucose molecules linked by β-glycosidic bonds) together with some pentosan (an arabinoxylan polymer) and a little protein.

These cells are stuffed with starch granules, which come in two sizes; large (about 15–20 μm diameter) and small (about 2 μm diameter). There are very many more small granules than large granules but they account for less than 5% of the weight of the starch. These starch granules are embedded in a matrix of hordein. This is an insoluble protein which provides a store of peptides and amino acids for the new plant. The whole of the starchy endosperm is surrounded by the aleurone, which is a triple layer of living cells.

The whole aim of the malting process is to get rid of as much as possible of the the β-glucan cell walls and some of the insoluble protein which would otherwise restrict access of enzymes to the starch granules. At the same time enzymes are developed which will, in the brewhouse, convert the starch into soluble sugars.

In the maltings the barley is steeped to raise the water content from 12% to around 45%. This process takes about 48 hours and consists of two or three periods when the grain is totally immersed in water, inter-

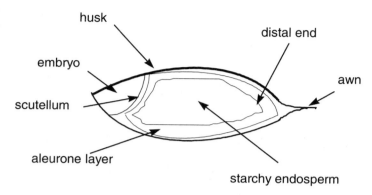

Figure 1.1 *Diagrammatic view (longitudinal section) of a barley grain*

spersed with 'air rests' when the water is drained off and fresh humidified air is blown through the grain bed to provide oxygen. The increased water content stimulates respiration in the embryo and hydrates the stores of starch in the endosperm. As the embryo activity increases, gibberellins are produced. These are natural plant hormones that diffuse into the aleurone, where they stimulate the production of hydrolytic enzymes during germination.

The moist grain is then allowed to germinate for a few days. During this time cool humidified air is again blown through the grain bed to keep the temperature down to around 16 °C and to stop the grain drying out. As gibberellins diffuse into the endosperm from the embryo they stimulate the aleurone cells to produce hydrolytic enzymes. These include amylolytic enzymes, which break down starch, proteolytic enzymes, which attack the protein, and cellulytic enzymes, which break down cell walls. Proteolytic enzymes include carboxypeptidases, which release one amino acid at a time starting from the carboxyl end of an amino acid chain, and endopeptidases, which can break peptide bonds in the centre of long amino acid chains. They can therefore very rapidly reduce the size of a protein or polypeptide. Next β-glucanases are produced. These break down the endosperm cell walls, making it easier for the other enzymes to diffuse out into the starchy endosperm. Last, but not by any means least, amylolytic enzymes are produced. The two most important are α-amylase and β-amylase, both of which can break down α-1,4 bonds. A debranching enzyme, which can attack the 1,6 bonds, is also produced, but this enzyme is quite sensitive to heat and so is normally inactivated during malt kilning.

All of these enzymes must diffuse into the starchy endosperm and begin the process of breaking down the cellular structure (the cell walls) and the

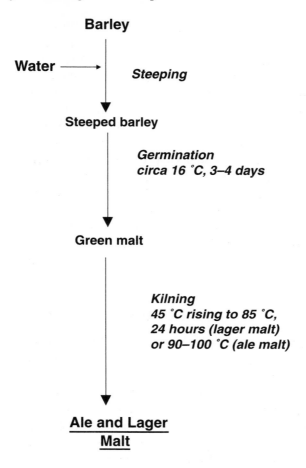

Figure 1.2 *Simplified flow diagram of the malting process*

stores of protein, starch and lipid in order to provide nutrients for the new plant. This process is strictly controlled by the maltster, who curtails it after four or five days. By this time most of the cell walls should have been digested, since if these are allowed to remain they will cause processing difficulties at a later stage. Part of the high molecular weight, insoluble protein will also have been broken down into smaller fragments (peptides and amino acids) and sufficient amylolytic enzymes will have been synthesised. Most of the starch remains intact, except for the small granules, which are the first to be digested during malting. If these small granules persist in the malt they can cause filtration problems for the brewer during the later stages of beer production.

The damp 'green' malt is dried in a kiln to prevent further enzyme activity and to produce a stable material which can be safely stored until needed for brewing. The kilning process also removes the volatile components responsible for undesirable 'grainy' flavours, and encourages the development of more attractive malty, biscuity flavours. This flavour development depends very much upon temperature and thus can be controlled by the maltster in order to produce a wide range of malts. The majority of commercial malts are fairly lightly kilned (up to 85 °C) in order to produce lager malts. In the UK a substantial proportion is kilned to a higher temperature (usually 90–100 °C) to give somewhat darker and more flavoursome pale ale malts. Higher temperatures (up to 200 °C) are used to produce speciality malts with flavours ranging from the toffee, caramel flavours of crystal malts to the sharp astringent flavours of roasted malts. These different malts can then be used by the brewer to produce beers with a wide range of flavours and colours (see Chapter 3).

MASHING

The brewing process converts the malt starch first to soluble sugars, then uses yeast to ferment these to alcohol. At the same time proteins are broken down into amino acids which can be used by the yeast as nutrients, coincidentally producing characteristic flavour compounds.

In the brewhouse the malt is crushed in a mill. Often a roller mill will be used – this keeps the husk largely intact so that it can serve as an aid to filtration later in processing. The crushed malt ('grist') is mixed with hot water in the mash tun and the whole mash is held at around 65 °C for about one hour. This temperature is chosen as it is the temperature at which malt (*i.e.* barley) starch will gelatinise – making it more susceptible to enzyme attack.

Sometimes other cereals ('adjuncts') may form part of the grist, in order to provide specific qualities in the beer. For example, small quantities of wheat are often used in ales to enhance the beer foam, while unmalted rice and maize grits may be used to improve the flavour stability of light-flavoured lagers. The more intensely kilned malts (crystal, amber, or brown malts) are used to provide colour and flavour in traditional British ales, while roasted malt and barley are used in the darker porters and stouts (see Chapter 3).

Like barley starch, wheat starch also gelatinises at 65 °C. Rice and maize starches gelatinise at higher temperatures, so if either of these cereals is used as an adjunct, it must be pre-cooked in a separate vessel (known as a cereal cooker) before being added to the mash.

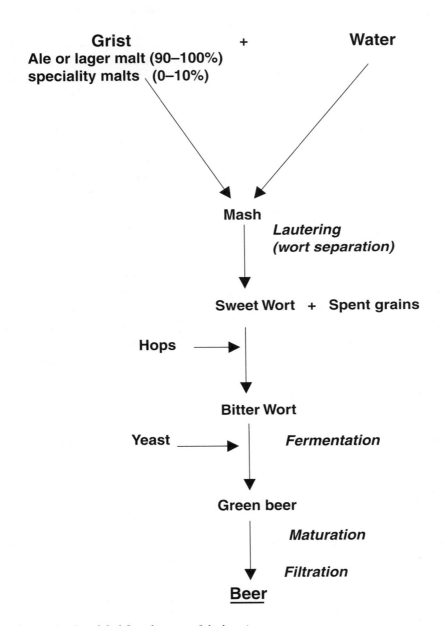

Figure 1.3 *Simplified flow diagram of the brewing process*

In some mashing systems, particularly those used for lagers, where the malts may have been less completely modified during malting, the mash may initially be held at a lower temperature (around 45 °C) to allow the breakdown of cell walls and protein which commenced in malting to continue. After about 30 minutes the temperature is then raised to 70 °C. At this temperature the starch will gelatinise and it can then be broken down by the amylase enzymes in the mash.

During mashing the amylolytic enzymes in the malt break down the starch into fermentable sugars. Cereal starch consists of approximately 75% amylopectin and 25% amylose. Amylopectin is a very large, branched molecule (the molecular weight has been estimated at several million) made up of glucose units linked by α-1,4 bonds (which give linear chains) and α-1,6 bonds (which give branch points). On average, each branch is made up of around 25 glucose units. Amylose, on the other hand, is a linear molecule made up of up to 2000 glucose units linked by α-1,4 bonds only (Figure 1.4).

Both α-amylase and β-amylase can hydrolyse α-1,4 bonds. β-Amylase attacks from the outer reducing ends of the amylopectin and amylose molecules, releasing free maltose (two glucose units), but stopping when it reaches an α-1,6 bond. In contrast, α-amylase attacks lengths of α-1,4 chains between branch points, releasing smaller, branched dextrins with long straight side-chains. These provide more substrates for β-amylase action. α- and β-Amylase acting together reduce amylose to maltose, maltotriose and glucose, but amylopectin gives rise, in addition, to many small branched dextrins which cannot be further broken down during mashing.

Thus after the conversion stage (mashing) a sweet syrupy liquid known as 'wort' is produced. This liquid contains mainly maltose and glucose, which are fermentable, together with significant quantities of small branched dextrins, which are not fermentable. There may be traces of larger straight-chain dextrins, the amount of which depends upon the enzymatic activity of the malt and the mashing conditions, and thus can to some extent be manipulated by the brewer. No starch should survive the mashing stage. The wort will also contain soluble protein, polypeptides and amino acids.

In traditional British ale mashing, the wort is separated from the spent grist in the mash tun by being allowed to filter through the spent grain bed into the next vessel. Hot water (usually at least 70 °C) is sprayed onto the top of the grain bed in order to extract and wash out the soluble components. This is known as sparging. A more usual practice nowadays is for the whole mash to be transferred to a separate vessel, the lauter tun. This vessel has a perforated base plate which allows the wort to run

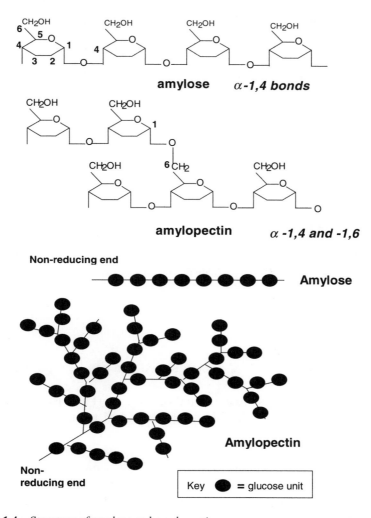

Figure 1.4 *Structure of amylose and amylopectin*

through into the next vessel, the kettle or copper, leaving the insoluble remains of the malt, (the spent grains) behind in the lauter tun.

WORT BOILING

In the kettle, hops or hop extracts are added and the wort is boiled quite vigorously. This has three effects:

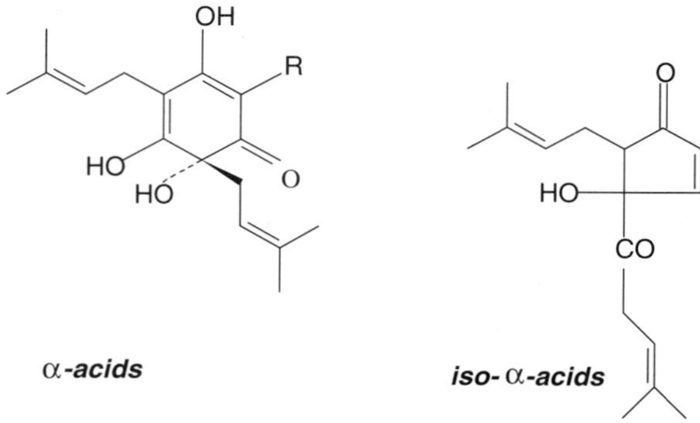

α-*acids*　　　　　　　　*iso-* α-*acids*

Acyl side chain (R)

——COCH$_2$CH(CH$_3$)$_2$ = Humulone

—— COCH(CH$_3$)$_2$ = Cohumulone

——COCH(CH$_3$)CH$_2$CH$_3$ = Adhumulone

Figure 1.5　*Structure of hop acids*

- The wort is sterilised.
- Much of the soluble protein is coagulated and can be separated off as the 'trub'.
- The α-acids in the hops are extracted into the wort and isomerised into iso-α-acids, which provide the characteristic bitter taste of beer (see Figure 1.5).

In addition to α-acids, hops contain essential oils, which contribute to the hoppy, floral and spicy aromas in beer (see Chapter 3). Most of these compounds are volatile and can therefore be lost by evaporation during boiling. In order to effect a suitable compromise between sufficient boiling to coagulate protein and isomerise the hop acids, but still retain the desired quantity of aroma compounds, the brewer may add part of the hop recipe part-way through the boil.

Also during boiling, browning reactions take place between the reducing sugars and the primary amines (particularly amino acids) in the wort, resulting in an increase in wort colour and some loss of free amino nitrogen. Browning reactions are complex and still not completely characterised, but basically consist of condensation reactions between simple sugars, such as glucose, with primary amines (for example the amino acid glycine) to give aldosylamines. These are relatively unstable compounds and can undergo Amadori rearrangement to form ketosamines, which condense with another aldose molecule to form diketosamines. Reacting in the enol form, these diketosamines can undergo further condensation reactions with additional amines to form a mixture of reddish-brown pigments, most of which contain a furfural ring. A simplified reaction pathway is shown in Figure 1.6. To some extent the amino acids act as catalysts, and the increase in colour is much greater than the loss of amino acids.

WORT CLARIFICATION

After boiling, the coagulated protein or 'trub', together with the spent hops, must be removed. Traditionally this was achieved by filtering the wort through the bed of spent hop cones. In modern breweries most of the hops are in the form of pellets or extracts, with much less waste leafy material to form a filter bed, and a vessel known as a whirlpool is used instead. The hot bitter wort is pumped into the whirlpool tangentially. The resulting swirling motion causes the trub to collect at the centre of the vessel as a conical mound. The clear wort can be removed from an exit pipe, which is situated to the side of the vessel. The bitter wort is then cooled to fermentation temperature by passing it through a paraflow heat exchanger.

FERMENTATION

Fermentation takes place at 7–13 °C for lagers or 16–18 °C for ales. Yeast is mixed with the cooled wort and the mixture pumped into the fermenting vessel. During fermentation the yeast takes up amino acids and sugars from the wort. The sugars are metabolised, with carbon dioxide and ethyl alcohol being produced under the anaerobic conditions found in brewery fermentations (Figure 1.7):

$$C_6H_{12}O_6 \rightarrow 2CO_2 + 2C_2H_5OH \qquad (1.1)$$

The amino acids are used for cell growth, so that at the end of fermentation the yeast will typically have increased its mass by up to 10-fold. The yeast also produces a number of flavour-active volatile compounds, mainly higher alcohols and esters, the exact profile of which will vary from strain to strain. (More details of the contribution of yeast to beer flavour are given in Chapter 3.) Thus the yeast is responsible for much of the unique character which distinguishes one beer from another. Most brewers have their own strain or strains, which may have been in use since the brewery was founded, decades or even centuries ago.

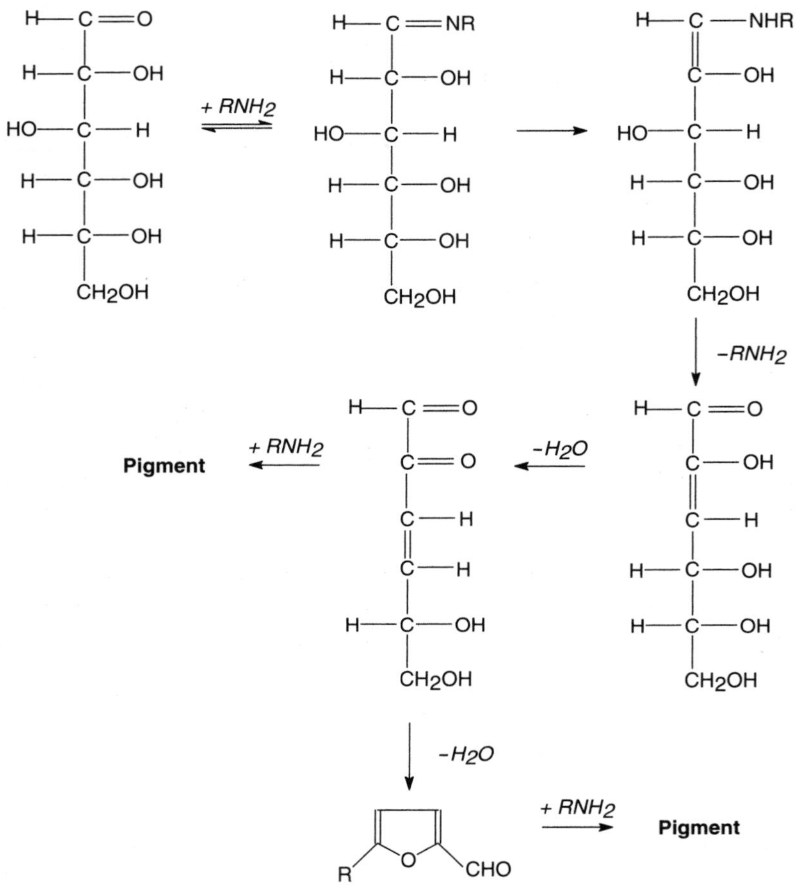

Figure 1.6 *Simplified non-enzymatic browning reactions in wort boiling*

Once the yeast has fermented all the available sugars, metabolism slows down, and with it, formation of carbon dioxide and ethanol. The yeast cells flocculate together to form clumps, which may either drop to the bottom of the vessel or rise to the top and float on the surface of the liquid. In general, lager strains are bottom-fermenting while ale strains are top-fermenting. Traditionally, therefore, ales were fermented in open vessels and the yeast head skimmed off the top at the end of fermentation. Nowadays, however, both ales and lagers are frequently fermented in closed cylindroconical vessels. When the fermentation has ceased, the vessel is cooled to 0 °C, which causes both types of yeast to drop to the

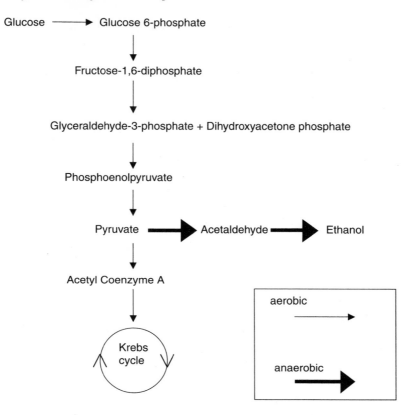

Figure 1.7 *Conversion of glucose to ethanol in yeast*

bottom. The bulk of the yeast can then be separated from the fresh beer in a process known as 'racking'.

MATURATION

This freshly produced or 'green' beer still contains undesirable flavour compounds and these must be removed by conditioning. During this time the relatively small proportion of yeast which remains in contact with the beer has two effects. Firstly, more carbon dioxide is produced – this carbonates the beer and purges it of unwanted volatile compounds. Secondly, the yeast chemically removes certain other flavour-active components. In particular it catalyses the reduction of flavour-active vicinal diketones such as diacetyl, to diols, which are not flavour-active (see Chapter 3). It is important that this reaction should proceed to completion during conditioning, since diacetyl and other vicinal diketones have very low flavour thresholds and can impart distinct flavours, typically described as butterscotch. Such flavours can be an essential element of, for example, some red wines and, to a lesser extent, some ales, but are undesirable in lighter ales and lagers. Traditionally this conditioning period for lagers is extended for several months – indeed the word lager comes from the German word meaning 'to store', with the beer being stored underground in cool limestone caves long before refrigeration was invented. In more recent years, however, procedures have been devised whereby, for most beers, this conditioning period can be decreased to days rather than weeks. For example, a short period of storage at a slightly higher temperature, around 12 °C, will enhance the formation and subsequent breakdown of diacetyl. For this reason such treatment is often known as a 'diacetyl rest'.

PACKAGING

After conditioning, the beer may be centrifuged to remove the remaining yeast, then chilled, filtered and packaged in bottles, cans or kegs. This is described as brewery-conditioned beer and represents most of the beer on the market today.

In the UK traditional cask-conditioned beers were racked directly into wooden casks, together with a small amount of yeast and isinglass finings to promote clarification. The solubilised collagen in finings has both positive and negative charges, but at the pH of beer their overall charge is positive. Thus they react readily with yeast cells (whose overall charge is negative) and with negatively charged proteins. They will also react with positively charged proteins, but to a lesser extent. The resulting large aggregates of particles fall to the bottom of the cask. Such casks are generally kept in the brewery for only a short time, often less than seven days, before being transported directly to the public house (or other retail venue), where they are put on stillage. The cask is placed in a horizontal position in a cool cellar without moving to allow any sediment that has accumulated from the finings, yeast and protein to fall to the bottom of the cask, allowing clear beer to be run off from above. Such casks need careful and expert handling in order to provide bright clear beer and there are always great losses due to the beer being entrained with the sediments. Possibly as a consequence of these disadvantages, cask beer has declined significantly as a proportion of the UK market in recent years.

SUMMARY

To summarise, beer is made from an aqueous extract of barley grains which have been allowed to germinate. Enzymes produced during germination digest the cereal starch to form sugars and these are then converted into alcohol by yeast. Hops are added to provide characteristic flavours and aromas.

FURTHER READING

1. D.E. Briggs, J.S. Hough, R. Stevens and T.W. Young, *Malting and Brewing Science*, Volumes 1 and 2, Chapman and Hall, London. Reprinted 1986.
2. C. Bamforth, *Beer. Tap into the Art and Science of Brewing*, Insight Books, Plenum Press, 1998.
3. J.S. Hough, *The Biotechnology of Malting and Brewing*, Cambridge University Press, 1985.
4. I.S. Hornsey, *Brewing*, Royal Society of Chemistry, Cambridge, 1999.
5. G. Fix, *Principles of Brewing Science*, Brewers Publications, Colorado, USA, 1989.

Chapter 2

Beer Quality and the Importance of Visual Cues

INTRODUCTION

When a consumer is presented with beer in a glass, he is immediately aware of not only the glass but also three facets of beer quality: its foam, colour and clarity. Each of these parameters is important in its own right, and can influence a consumer's future choice of product or, in extreme circumstances, result in the consumer returning the product untouched. In this chapter, the determinants of foam, colour and clarity will be discussed in turn. The main focus will be on foam and colour as a haze-free or bright beer is generally mandatory for consumer acceptance. The presence of haze is often a result of the deterioration of beer with time and so is discussed in Chapter 4.

A last point about consumer appraisal of foam, colour and clarity is that, with the current trend of consumers to drink straight from the can or bottle, the visual impact of the product itself can be secondary to the appearance of the package. Indeed, the importance of the package cannot be underestimated: the protection of beer from light plays a key role in maintaining the flavour integrity of beer, but there is still great consumer demand for products packaged in non-protective green or clear glass bottles. This issue will be discussed in more detail in Chapter 3.

PHYSICAL PROPERTIES OF BEER FOAM

What is Beer Foam?

Foams are colloidal systems comprising of a discontinuous gaseous phase and a continuous liquid or solid phase. The amount of liquid held up in a beer foam is time dependent, with a more or less wet foam rapidly draining to leave an essentially solid network of bubble walls. These walls are deposited from the liquid phase, a process which begins immediately upon the formation of the bubbles. The process of foam drying is often termed drainage, as generally the liquid leaves the foam under the influence of gravity. Indeed, many foam measurements are based on liquid drainage, not least because measurement of drained liquid volumes over a specified time period is relatively straightforward.

However, drainage is not the only parameter by which beer foam can be judged. From a consumer point of view, drainage results in a modest change of foam volume and may not be as apparent as bubble coalescence – the combination of two or more bubbles to form fewer, larger bubbles; or disproportionation – where the larger bubbles increase in volume at the expense of smaller bubbles. These latter two can be readily perceived visually, as they give a coarser, less aesthetically pleasing foam. For many beer foams, drainage precedes coalescence and disproportionation, so that the determination of drainage essentially reflects the early lifetime of beer foam.

Nucleation

Nucleation is a term used to describe the process of bubble formation. Bubbles may be generated in beer by either dispersal or condensation methods. The former involves the direct injection of gas into beer, and the latter is brought about by inducing the discontinuous gas phase to agglomerate from some simpler (*e.g.* solvated) state.

Dispersal Methods

The simplest dispersal system to consider is the injection of gas *via* a capillary. The surface energy of the growing bubble will be minimised if it takes up a pseudo-spherical geometry. For a spherical bubble forming on top of a capillary of circular cross-section, the bubble will be released when its buoyancy is greater than the surface tension effects of the bubble adhering to the perimeter of the top of the capillary.

Bubble volume is proportional to surface tension and inversely proportional to liquid density. By analogy, foaming *via* a sinter, which is essentially a heterogeneous collection of pores, generates a range of bubble sizes. This has been shown to have implications for the ultimate stability of foams, as a heterogeneous bubble size distribution is in principle more prone to disproportionation and therefore more rapid foam breakdown.

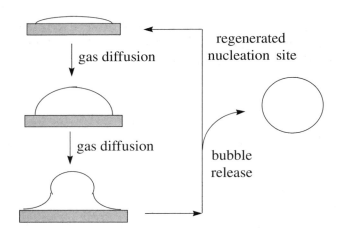

Figure 2.1   *Bubble formation at a nucleation site. Initially, small pockets of gas grow by the diffusion of solvated gas. As the bubble grows, it experiences increasing buoyancy, and detaches when this buoyancy exceeds the surface tension effects of the nucleating interface. Not all of the gas is detached as a bubble, which enables the process to repeat itself*

Condensation Methods

Condensation methods of bubble formation can either be homogeneous or heterogeneous. The former is unlikely to occur on energetic grounds but is observed, for instance, on removing the crown cork from a glass bottle (beer at 5 °C) where the temperature can drop to about −36 °C due to rapid gas expansion.[1] Much more likely is heterogeneous nucleation. Here, gas already present (usually air in the case of beer dispense), is expanded by the diffusion of gas (either carbon dioxide or nitrogen) from solution into the gas phase. As the bubble grows, it experiences greater buoyancy, until it breaks away from its nucleating surface, leaving a small pocket of residual gas to begin the process again (Figure 2.1).

The use of small particles to induce rapid foaming, rather than providing rough surfaces *per se* on which bubbles can form, has been shown to be effective only when they entrap air pockets and are, in fact, relatively ineffective when totally wetted. The effectiveness of etched glasses for foam generation and replenishment has been recognised for some time. Nevertheless, it is likely that such glasses, which rely on air trapped in the etched part of the glass during dispense for bubble nucleation, may lose their efficacy because of the difficulty in maintaining the cleanliness of the etched moiety. An alternative route for inducing bubble formation is by cavitation. This is a process whereby nucleation sites are generated by agitation of the beer, resulting in the instantaneous separation of beer and vessel. Gas rapidly diffuses into these vacua, and the process of bubble growth and detachment can begin. Ultrasound is a highly potent cavitation method and can result in very rapid, uncontrollable gushing from a bottle or can.

Foam Ageing

Once a bubble has detached, buoyancy (and, to a lesser extent, opposing drag forces) dictates that the bubble rises through the body of liquid which forms its environment. Consideration of physico-chemical parameters allows the time taken for a bubble to travel through a medium to be calculated. Presumably, foam-active species concentrate in the bubble walls as the bubble moves, lowering the surface tension of the bubble and therefore its energy. Alternatively, it is possible that the rate of bubble growth at the site of nucleation influences the maturity of the bubble on release – *i.e.* the surface tension may have already approached a minimal value. One area which has not been studied in detail is the dynamics of adsorption of foam-active species into bubble walls. The rate of bubble formation is likely to be of critical importance, essentially limited by the inherent foamability of the endogenous protein present. As will be discussed below, hydrophobic interactions are essential to beer foam integrity, and species which disrupt these interactions, such as competing surfactants and chaotropic reagents are potentially damaging.

The bubble will grow as gas diffuses into it and, of course, the top pressure on the bubble reduces as the bubble travels upwards to the beer–air/foam interface. Once the bubble reaches the interface, it is initially (to a first approximation) spherical. Nevertheless, liquid drainage from between these independent bubbles soon occurs, so that lamellae soon form between bubbles. This is apparent as a transition from 'wet' to 'dry' foam. The diffusion of gas from smaller bubbles to larger bubbles, along a pressure gradient, results in a process called disproportionation. Eventually, bubble lamellae rupture to give fewer, larger bubbles (coalescence). Bubbles exposed to the foam–air interface can also lose gas rapidly along a pressure and concentration gradient. This is especially apparent for carbon dioxide bubbles, where the concentration gradient is large, and the gas is readily soluble in the bubble lamellae. The result is a small, low pressure bubble, with an excess of bubble wall material present.

In summary then, during the process of foam ageing, bubbles move upwards and away from the beer–foam interface, as younger bubbles arrive beneath those already there. The resultant effect is a crude stratifi-cation of beer foam, with spherical bubbled, wet foam at the bottom adjacent to the beer–foam interface, and a pseudo-polyhedric bubbled foam above. Bubble size distribution as a function of vertical displace-ment is difficult to predict and is a result of the relative rates of coales-cence and disproportionation, as well as rapid bubble shrinkage at the foam–air interface.

BEER FOAM COMPONENTS

Foams are inherently difficult to study. This is not only because of heterogeneity – foams may consist of two or arguably even three phases – but also, in the case of beer foam, because it is transient and hence study is essentially restricted to the observation of a dynamic system. Beer foam is stabilised by the presence of beer polypeptides and hop bitter acids, but a number of other beer components can also substantially affect beer foam and are described below.

Proteins/Polypeptides

The heterogeneity of foam proteins has meant that detailed characterisa-tion has proved difficult. Kaersgaard and Hejgaard[2] detected four major antigens in beer, the major one originating from protein Z in barley ($M_r \sim 40 \text{kDa}$). There was also a significant proportion of antigen which was derived from yeast cells. It has been suggested that proteins of different, specific sizes are responsible for beer foam stability. Others report that specific groups of proteins are important for imparting stabil-ity.[3] Thus non-enzymically glycosylated proteins, glycoproteins, proteins with high isoelectric points, or hydrophobic proteins have variously been proposed as being key contributors to beer foam stability. In support of this latter point, it has been shown that various protein fractions from beer foam, isolated on the basis of their hydrophobicity, were found to correlate strongly with foam stability. In addition, all of these isolated protein fractions bore components of $M_r \sim 40 \text{kDa}$, suggesting that it is the way in which protein structures are modified during malting and beer production which affects beer foam stability rather than molecular weight *per se*.

More accurately, it is amphipathicity (*i.e.* the presence of both polar and non-polar regions on the same molecule) rather than hydrophobicity of foaming proteins that is the crucial property for foam activity. This means that, at the air–liquid interface, the hydrophobic regions can extend into the gaseous phase, whilst the hydrophilic portions can extend into the polar aqueous phase. One model for the aggregation of proteins at interfaces invokes protein denaturation at the interface as a prerequisite for foam formation. This is feasible on energetic grounds: a protein's most stable (*i.e.* lowest energy) conformation in an aqueous medium, which would involve minimal exposure of hydrophobic regions to water, will not be the lowest energy conformation at an air–liquid interface. Nevertheless, the kinetics of protein denaturation are dependent on the tertiary structure of the protein. Thus β-casein, a relatively unstructured protein, will more quickly attain its lowest energy conformation at the interface than more rigidly structured species such as lysozyme. Comparative studies of these two proteins have led to the suggestion that foamability is dependent upon the rate of protein denaturation.

In addition to the structural amphipathic polypeptides, it has become apparent that certain proteins can protect beer from lipid damage. Clark *et al.*[4] identified wheat-derived proteins which can protect beer foam from lipid destabilisation, which they contend could be due to lipid–protein interactions. These so-called lipid binding proteins have also been identified in barley. A quite distinct material, lipid transfer protein (LTP1) was isolated from a foaming tower. This 10 kDa protein shows excellent foamability in the presence of higher molecular weight proteins, although it demonstrates little foam activity itself. The activity of LTP1 isolated from barley is much lower.[5]

Polysaccharides

Polysaccharides are polar and are not particularly foam-active in themselves. However, when present as glycoproteins, they provide large polar regions which can extend deep into the bubble lamellae and Plateau borders. There is evidence to suggest that the carbohydrate moiety of beer foam glycoproteins is not dissimilar from amylopectin. Not only could carbohydrates result in increased localised viscosity, slowing down liquid drainage, but they could also thicken the electric double layer across lamellae, giving rise to thicker lamellae. Alternatively, it has also been proposed that polysaccharides can help to cross-link species in adjacent interfaces as they are brought into close proximity by liquid drainage.

Hop Bitter Acids

Iso-α-acids, the major source of bitter flavour in beers,[6] are concentrated in beer foam. Various workers have found that the isocohumulones were concentrated to a lesser extent than their less polar isohumulone and

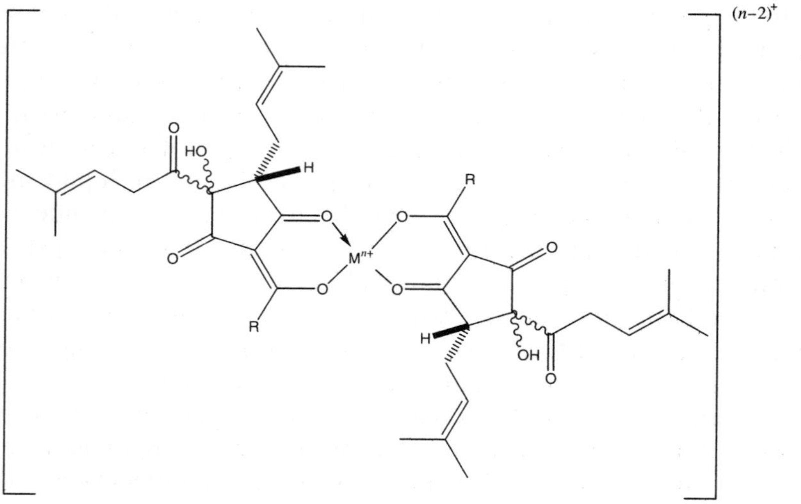

Figure 2.2 *Possible structure of iso-α-acids chelated to a metal cation. Cations with a charge n > 1 are prone to precipitate iso-α-acids, presumably due to the shielding of part of the polar β-triketone moiety. The exposure of the side-chains should be conducive to their participation in hydrophobic interactions*

isoadhumulone counterparts, and one report has indicated a seven-fold concentration of *trans*-isoadhumulone in beer foam relative to the remaining liquid beer. This suggests that it is the hydrophobicity of the hop acids which influences their partitioning into beer foam. The role of iso-α-acids in the stabilisation of beer foams is discussed later in the chapter.

Metal Cations

Metal cations are well-known to induce beer foaming. Iron salts were used in the past to improve beer foam, until their role in flavour deterioration was recognised. Indeed in the 1950s Rudin demonstrated an excellent correlation between foam stability (as measured using Rudin's eponymous foam drainage technique) and iso-α-acids content of beers to which had been added nickel(II) at a level of $10\,mg\,l^{-1}$. There is much evidence to suggest that iso-α-acids interact with metal cations, presumably due to the β-triketone moiety of the hop acids which has an affinity for polyvalent metal cations (Figure 2.2). Iso-α-acids have also been shown to bind potassium ions in a cooperative manner in the presence of a range of polyvalent cations. Such binding could have consequences for the ability of hop bitter acids to stabilise beer foam structure.

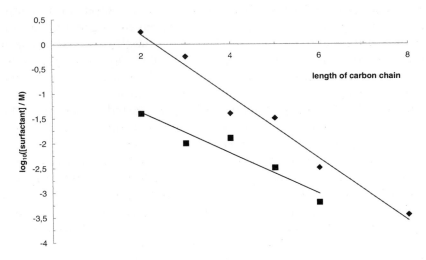

Figure 2.3 *Ferguson plot for the effect of straight-chain alcohols and acetate esters on Ross & Clark foam measurements. The concentrations are those required to completely destroy foaming. Linearity here suggests that the effects are general, rather than specific to a given side-chain (Based on data from Lienert[7]). Note that whilst ethanol is weakly surface-active, its concentration in beer (about 0.5–1.2 M) is such that it is likely to be challenging to beer foam stability* (Courtesy of the European Brewery Convention)

Alcohols and Lipids

The presence of ethanol increases the viscosity of water so that the presence of ethanol would be expected to result in a reduction in the rate of liquid drainage from a foam. Lipids and higher alcohols are both examples of amphiphathic species which can, therefore, adsorb into water–gas interfaces. The addition of either to beer is disastrous for beer foam. Lienert[7] showed that, for a series of straight-chain alcohols and acetate esters, the effect is not specific for a given molecule, and that an increase in the hydrophobic chain-length in turn leads to a more potent beer foam destabiliser (Figure 2.3). The fact that beer foam formation and stability can be recovered if the lipid is left in contact with the liquid beer for a sufficient length of time suggests that the lipid is neutralised in some way. The presence of lipid-binding proteins in barley and wheat suggests that these species may help to protect beer from lipid damage.

Gas Composition

Traditionally, the only significant gas component of beer foams was carbon dioxide. Guinness first introduced Draught Guinness as it is known today in 1964. This was facilitated by the development by Guinness of the two-part keg – one part for mixed gas (nitrogen and carbon dioxide) and the other part for the beer. It is the presence of nitrogen which accounts for the distinctive appearance of Guinness stout, with its tight, creamy head. This creaminess is due to the much smaller bubble sizes which nitrogen gas can generate relative to carbon dioxide.

In the late 1980s, so-called 'widget' products started to appear in the UK and soon began to enjoy substantial popularity amongst consumers. The value of this market is reflected in the sheer number of patents which exist for widgets – 72 up to May 1996 excluding duplicates. These widgets introduce a cavity into small-pack products whereby liquid nitrogen, introduced into the beer just before can closure, is allowed to diffuse into the widget. As nitrogen gas is poorly soluble in beer, the addition of liquid nitrogen results in a high internal pressure within the can. When opened, this pressure is released and the nitrogen gas trapped within the widget at the bottom of the can quickly forces its way through the bulk liquid, generating substantial amounts of foam. Indeed, a widgeted can of beer can provide a very practical demonstration of the temperature dependence of gas solubility – a chilled widgeted can foams in a much more controlled manner than a can which is not chilled. A substantial number of carpet stains are a testament to this phenomenon!

pH

The pH of beer has been shown to be a significant factor for beer foam stability. Melm *et al.*[8] found that multiple regression models for foam stability were well-modelled when pH was a variable, with lower pH values giving higher foam stabilities. It is important to remember that pH is a measure of proton activity in aqueous systems, so that attempts to measure foam 'pH' values are inherently flawed. Why a lower beer pH should result in greater foam stability is not clear. A significantly greater proportion of the hop bitter acids are undissociated at the lower range of beer pH values. This in turn means that the hop acids are more hydrophobic and can therefore adsorb into the interface more efficiently. Note though that they will still bear the polar β-triketone system and therefore retain a degree of amphipathicity.

Other Components

Polyphenols

Beers contain a range of monomeric and condensed polyphenol structures. They interact strongly with proteins, and this effect is manifested in a diverse number of observations (*e.g.* leather tanning, sensory astringency). However, there appears to be little preferential adsorption of total polyphenols into beer foams. Crompton and Hegarty[9] speculated that high molecular weight polyphenols covalently cross-link polypeptides in foam lamellae, whilst low molecular weight species (*e.g.* catechin) are ineffective.

Melanoidins

Melanoidins have been shown to slow down the rate at which liquid drains from foams on their addition to base beers, and help to protect such beers from lipid destabilisation. Furthermore, some hydrophobic, foam-stabilising fractions appear to contain significant quantities of melanoidins. Lusk *et al.*[10] found that melanoidins form stable foams, even in the absence of proteins.

FOAM PARAMETERS

Beer foam may be characterised by a number of measures. Many of these factors are interrelated, but it is useful to consider them separately. Of course, foam measurements are often influenced by more than one of these parameters.

Foamability

This is a property which indicates how readily a foam will form a solution. In terms of protein-stabilised foams, this has been described as the ability of proteins to denature at the liquid–gas interface. As already described, a relatively unstructured protein, such as β-casein, will foam much more readily than structurally more rigid substances such as lysozyme. Unfolding occurs with the hydrophobic portions of the protein looping into the gas phase and the polar regions remaining in the aqueous phase. It should be stressed that good foamability does not necessarily confer good foam stability. Thus, whilst lysozyme is difficult to foam, the foam once generated is stable. The foamability of beer foam is enhanced by the presence of heavy metal cations and a number of so-called 'gushing promoters'.

Foam Stability

Foam stability is the ability of a foam, once formed, to resist degradation processes. The exact definition of foam stability depends on which measure is employed (these are discussed below). The stability of beer foams is enhanced by hop bitter acids and their chemically-modified variants, and exogenous stabilisers such as propylene glycol alginate and pectins. Foam stability can be considered to be a function of the viscosity of the bulk liquid, surface viscosity, the Marangoni effect (see Foam Structure), and the repulsion of electric double layers (which helps to maintain the bubble lamellae intact).

Foam Drainage

This is a convenient parameter which can be used to assess foam stability. It is the rate at which liquid runs from the bubble lamellae and Plateau borders, and is primarily a function of the bulk liquid viscosity and the volume of liquid held up in the newly-generated foam.

Cling

Cling (alternatively lacing or foam adhesion) is the residual foam which adheres to the glass when the beer is removed. The importance of cling for positive consumer response has been noted previously.[11] Methods used for the measurement of cling are based either on assessing the total amount of material present, or the relative area of coverage of the glass. Nevertheless it should be remembered that aesthetically-pleasing lacing may only be represented by a small quantity of material.

Viscoelasticity

An adsorbed surface, when deformed, can either adapt to absorb the deformation or, alternatively, resist the deformation – stretching and then recovering when the applied stress is removed. The former case approximates to a surfactant-based adsorbed layer, characterised by highly mobile species which can readily diffuse into any regions of thinning. Clearly, such a layer will retain little memory of its former state. In contrast, foams stabilised by proteins are characterised by low rates of lateral diffusion and will resist deformation. The layer will therefore retain some memory of its former state and, to some extent, recover when the stress is removed. Measures of viscoelasticity, such as the dilational modulus, essentially reflect how much a foam can resist deformation. Studies carried out on commercial beers show that their dilational moduli are at the lower limit for protein-stabilised interfaces.

Lateral Diffusion

This is the rate at which a molecule adsorbed into the interface can diffuse across the surface. This is a function of the molecule itself, and is therefore a relative measure of a given species in a range of foaming systems. A rigid, essentially solid interface is characterised by low lateral diffusion rates, whilst mobile surfaces, such as those of purely surfactant foams, will have high lateral diffusion rates.

Film Thickness

The thickness of bubble lamellae is a property inherent in the composition of the liquid beer. Bubble walls thin as liquid drains from between them until they eventually lose their spherical symmetry. Adjacent bubble walls will approach until either there is rupture, or a minimal thickness (determined by electric double layer repulsions) is attained.

Bubble Size

Generally, dispensed beer contains bubbles of a range of sizes. For a given beer, the presence of nitrogen will afford a smaller mean bubble radius. As the internal pressure of a bubble is inversely proportional to its volume, a heterogeneous bubble size distribution means that there are pressure (and thus energy) gradients within foams which increases the inherent instability of beer foams. Bubble size can play an important part in the results of foam measurement methods, particularly when sinters are employed. This problem has recently been addressed for foams generated for measurements using the NIBEM foam analyser.[12] The colour of the foam is to some extent dependent upon bubble size/bubble density, being reliant upon the radii of film curvatures and their thicknesses.

FOAM STRUCTURE

Several alternative structures have been proposed for beer foam. One model postulates that negatively-charged hop bitter acids interact with positively-charged protein residues (*e.g.* protonated lysine side-chains). Whilst this can help to explain why iso-α-acids are concentrated in beer foams, it does not account for the foam-enhancing properties of heavy metal cations or why hydrophobic species (lipids, higher alcohols) are so damaging to beer foam stability.

More likely is that the amphipathicity of both proteins and hop bitter acids is essential for maximal beer foam stability. Thus, whilst polar interactions (dipole–dipole, ionic, hydrogen-bonding) can hold together the beer foam structure, the stability relies on the presence of hydrophobic interactions. It is difficult at the moment to divorce the efficacy of adsorption of materials into bubble walls and the ability of these materials to stabilise foams on a molar basis. One possible suggestion is that foam formation is selective, and that the most foam-active materials will be adsorbed into foams to the greatest extent.

The iso-α-acids interact with polyvalent metal cations, an effect which can be monitored by UV/vis spectrophotometry. Thus, it is possible that the stabilising properties of heavy metal cations is by their effective chelation to hop bitter acids. Incidentally, this increases the hydrophobicity of the hop bitter acids, as seen from their reduction in solubility in aqueous media.

The exact role of hydrophobic interactions in the stability of beer foams is unclear. Presumably, the relatively low water content of mature foams is further ordered by the presence of hydrophobic functionalities. Energetically, this is unfavourable because of the accompanying decrease in entropy, but this is dependent upon the absolute amount of water bound up in the foam matrix. There may not be an insurmountable energy barrier as much water will be bound up with the hydrophilic portions of the proteins and polysaccharides present. As mentioned above, the dynamics of bubble formation will dictate whether or not chaotropic species present in beer will cause a foam problem.

The effect of lipids and higher alcohols on beer foam can only be described as disastrous. Bearing in mind that protein-stabilised foams are rigid and characterised by low rates of lateral diffusion, the presence of these organic components interrupts hydrophobic interactions and breaks up the longer-range surface interactions. Addition of surfactants to protein-stabilised systems decreases foam stability to a minimum before it begins to recover at higher surfactant concentrations. Here, the protein cannot adsorb to the interface because of the high surfactant concentrations. Thus, there are three distinct phases of foam structure which are illustrated in Figure 2.4. First, in the absence of surfactants, a rigid protein-stabilised interface exists. The addition of small quantities of surfactant destabilises the surface to a minimum value, whereby the free diffusion of surfactant into the surface is hindered by the protein present (and hence preventing restabilisation *via* the Marangoni effect), whilst the surfactant in turn interferes with cross-linking hydrophobic interactions of the depleted protein layer. Increasing the relative concentration of the surfactant results in effective exclusion of the protein from the interface, and free diffusion of the surfactant can now occur to stabilise layers in accordance with the Marangoni effect.

Surfactant only

Protein only

Mixed system

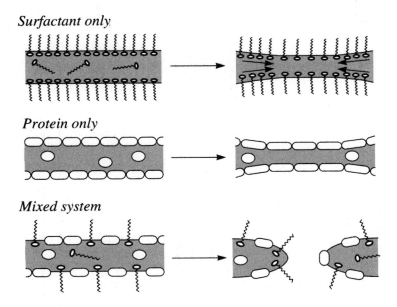

Figure 2.4 *Illustration of protein-stabilised, surfactant-stabilised, and mixed systems. The protein-stabilised layer is characterised by a rigid, low mobility layer which resists deformation. This is in contrast to the surfactant-stabilised layer, where applied stress, increasing the film surface area, is repaired by rapid diffusion of surfactant back into the weak spot (i.e. the Marangoni effect). In a mixed system, protein stabilisation is not possible because of the interference of the surfactant in the protein–protein interactions. Surfactant stabilisation is minimal because of the steric hindrance of bulky protein residues, effectively preventing free diffusion of surfactant* (Courtesy of the European Brewery Convention)

IMPROVING FOAM STABILITY

Propylene Glycol Alginate (PGA)

This is a heterogeneous substance formed by the partial esterification of alginic acid with propylene oxide (Figure 2.5). The degree of esterification is important in determining the efficacy of foam stabilisation and colloidal stability of the final beer, and is typically 80–90%. The mechanism by which PGA stabilises foams is unclear, although it may be due to ionic interactions between the essentially dissociated PGA and the foaming proteins. This mechanism suggests that PGA could function by thickening the electric double layer of lamellae. PGA also appears to stabilise beer foams even in the face of lipid damage. Typical levels of application are $\sim 50\,\mathrm{mg\,l^{-1}}$. Of increasing importance though is to find alternatives

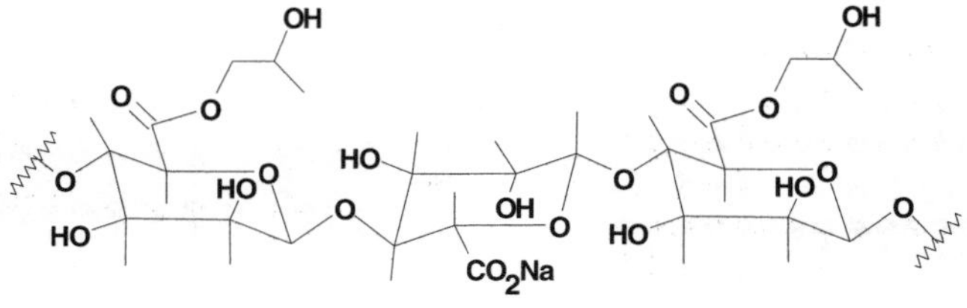

Figure 2.5 *Idealised structure of propylene glycol alginate (PGA). A high degree of esterification is essential if colloidal stability of the product is not to be compromised*

to PGA as labelling legislation is likely to require its listing as a beer ingredient in the future, detracting from the wholesomeness of the product.

Chemically-modified Iso-α-acids

The addition of four or six hydrogen atoms to iso-α-acids can give rise to tetra- and hexahydroiso-α-acids respectively. These species have a greater propensity to stabilise beer foams than their native counterparts as well as affording beer protection against the formation of lightstruck flavour and, on a molar basis, higher bitterness intensities. While there are no published pKa values for these substances, it is worth noting that the tetra- and hexahydroiso-α-acids are also less polar than iso-α-acids themselves, once again indicating the importance of hydrophobic interactions of the hop acids to the stabilisation of beer foams. Generally, reduced iso-α-acids will be added as late as possible during beer production, to minimise losses of these highly surface-active substances. Their more potent bitterness (particularly the tetrahydroiso-α-acids) effectively limits the quantity which can be added, but even at $3–5\,\mathrm{mg\,l^{-1}}$ they can effectively stabilise beer foams.

Choice of Raw Materials

The composition of the grist for beer production should be critical for the quality of the resultant foam as most of the protein found in foam comes from this source.[2] It is reputed that the use of wheat as part of the grist also results in improved beer foam stability. The effects of the variation in hop acid profile on foam quality is less clear, although one might expect there to be a difference due to their range of hydrophobicities.

Dispense Hardware and Gases

Whilst attention is correctly aimed towards improving and/or maintaining beer foam quality by careful consideration of the chemical composition of beer, the physical attributes of beer foam and their effect on foam quality should not be overlooked. Thus, the bubble size distribution of the foams generated, the temperature of the dispensed beer and the cleanliness of both dispense equipment and serving glasses are all critical in determining the final foam quality of beer as perceived by the consumer.

FOAM ASSESSMENT

There are several points to bear in mind when looking to assess beer foam. Firstly, the context of the measurement needs to be explicitly defined – is the measurement required to indicate that the foam is performing as is expected for a given product before it leaves the brewery? Or, alternatively, is there a problem with the beer foam at the point-of-sale? Does the foam performance need to be investigated to identify why performance has varied? All of these issues dictate different requirements of a measurement.

Foam quality control is carried out by brewers to ensure that, when sold, the foam will perform as expected by the consumer. As consumer assessments are prohibitively expensive to use on a routine basis and also prone to a high degree of variability, some form of instrumentation is almost invariably invoked to provide an estimate of performance in trade. One of the most commonly used methods is that known as the NIBEM (Figure 2.6). Here, beer in small-pack is forced into a cuvette, usually by carbon dioxide gas. The foam formed is then monitored for its rate of collapse. The longer this takes, the better the foam is considered to perform. Other methods, such as the Rudin test, require the use of degassed beer which is then foamed with either nitrogen or carbon dioxide and the rate at which liquid drainage subsequently occurs is used as a measure of beer foam stability.

Assessment based on human responses is potentially the most valuable and indeed is an approach taken by several brewing companies worldwide. Beer may be dispensed in real-time, or alternatively photographs or video images of the foam are presented to the assessors. The latter is likely to find increasing favour as sensory analysis becomes more computer-based, because of the ability to select a beer dispensed in a typical manner. One such test which has been used in-house by the author assesses the subjective term 'quality', as well as a visual assessment of stability and

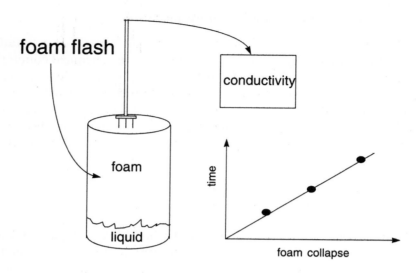

Figure 2.6 *The NIBEM foam stability test*

lacing quality. Whilst the data which can be generated from such tests is subject to a substantial amount of scatter, such methods can prove informative.

THE EFFECTS OF PROCESS ON FINAL FOAM STABILITY

The presence of iso-α-acids are essential for the foam quality of beer. Until recently, it was often therefore important that the utilisation of hop bitter acids was controlled. With the advent of downstream adjustment practices, the iso-α-acid content of beer should be controlled to an extent where this would not cause appreciable problems to the brewer.

The degree of malt modification can influence the molecular weight distribution of polypeptides in beer and therefore have an impact on final foam stability. Over-modification is likely to result in a poorer foaming beer, whilst malt that is slightly under-modified may be beneficial to beer foam. In this context, dilution of soluble nitrogen by use of high proportions of adjuncts can result in foaming protein, normally present in excess, becoming limited.

More difficult to control and identify are lipids and the presence of other surfactant materials. Small quantities of such substances can be highly damaging to final beer foam stability. Thus, even $2\,mg\,l^{-1}$ of linoleic acid in beer can result in a liquid which is difficult to foam. It is interesting that beer does seem to recover somewhat from lipid damage on storage, presumably as a result of the lipid becoming associated with lipid-binding proteins. Mineral oils are extremely detrimental to beer foam quality and their introduction into beer should be prevented wherever possible. Sources of adventitious oils include brewery gases, pump and compressor lubricants and canning components.

Foaming during beer production can lead to the irreversible loss of foam-active species. Thus foaming during fermentation results in the adsorption of hop acids into the yeast head. Laws *et al.*[13] suggested that these losses could be alleviated to some extent by folding the yeast head back into the fermenting wort, preferably when $> 1\%$ alcohol by volume has been reached. Irreversible denaturation of proteins, a process perhaps essential to the formation of beer foam, generally reduces protein solubility. Thus foaming will result in a loss of a proportion of the foam-active proteins from the bulk beer. Paradoxically, the use of silicone antifoams during fermentation is beneficial to final beer foam stability, provided that the antifoam is removed at filtration. This is on account of the suppression of foam head during fermentation, with attendant reduced losses of protein.

Attempts to improve the colloidal stability of beer by removal of haze-forming proteins (*e.g.* by the addition of silica hydrogels) can result in the simultaneous removal of foam-promoting proteins. Thus it is important that the use of these materials is suitably controlled.

BEER COLOUR

Perception of Colour

The perception of colour is a complex integration of the transmission, absorption and reflection of light. Visible light is characterised by its wavelength, which can vary continuously from around 400 to 700 nm. The eye does not respond to light as a broad spectrum of intensities. Rather, the human perception of colour is the result of three distinct signals sent from the eye to the brain. Cone cells at the centre of the retina respond individually to either green, red or blue light and it is these three responses that are interpreted by the brain as colour. Thus a distortion of the wavelengths of environmental 'white' light, by its passage through beer, confers colour (Figure 2.7). Colour is inherently descriptive, although from a scientific perspective a numerical description is preferable. However, care must be exercised as the human eye is not equally sensitive to all visible wavelengths, so any numerical appraisal of colour needs to take such biases into account. The colour of beer is a critical parameter

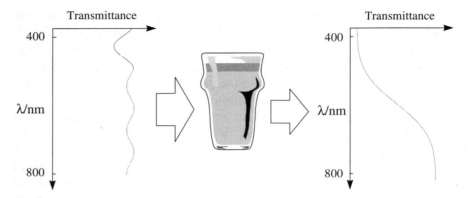

Figure 2.7   *Schematic representation of colour. The 'white' spectrum on the left is distorted by its passage through the beer sample, so that whilst the long (red) wavelengths pass through the short (blue) wavelengths are absorbed. Such absorption is due to light-absorbing molecules present in the beer*

for many consumers, as it allows instant classification of beer type – lager, ale, stout – and therefore requires careful control.

Light-absorbing Species in Beer

Melanoidins

Virtually all colour in beer forms during malting and wort production by four routes (Table 2.1). Firstly, the Maillard reactions (described in Chapter 4) ultimately yield complex mixtures of compounds loosely termed melanoidins. These are poorly defined, soluble pigments, with a range of colours from amber to yellow.[14] Given that they originate from thermal reactions between sugars and free amino acids, the protein content of the malt and the mashing regime will influence the quantity and profile of melanoidins that form. Maillard reactions can occur during germination and malt kilning, the roasting of barley for the production of stouts, mashing, lautering, wort boiling, trub separation, and wort cooling. The formation of melanoidins is favoured by elevated process temperatures.

Polyphenols

The second significant source of colour in beer is oxidised polyphenols. Generally red-brown in colour, they can be formed from polyphenols throughout the brewhouse. However other factors, such as malt milling and over-sparging, can influence the polyphenol yields. The role of oxy-

Table 2.1 *Main contributors to beer colour*

Component	Sources raw materials	process	Colour
Melanoidins	Malt, speciality malt	Wort boiling	Yellow, amber
Oxidised polyphenols	Malt, hops	Oxygenation, pasteurisation	Red, Brown
Iron, copper	Water, malt	—	—
Riboflavin	Malt, yeast	—	Yellow

gen then should not be underestimated. Thus aeration or oxygenation of wort whilst still hot will result in colour pick-up, and its presence in-pack will give elevated colour after pasteurisation. Approaches used to reduce beer polyphenol levels, such as PVPP adsorption (Chapter 4) can also reduce beer colour. Finally, very high temperatures used to produce highly coloured malts and roasted barley result in caramelisation reactions. These products have an intense red-brown colour and are used for ale and stout production.

Trace Metals

A third source of colour can be due to interactions with trace metals. Thus copper and iron can stimulate oxidation products (*e.g.* of polyphenols). Finally, although usually present at levels less than $1\,\mu M$, riboflavin can contribute significantly to the colour of pale lager-type beers.

Beer Colour Measurement

The form of a transmission spectrum of beer generally has few features (Figure 2.8). Transmission monotonically increases from around 350 to 800 nm, and is virtually transparent to light at wavelengths greater than 700 nm (*i.e.* into the infra-red region).[15] The simplest method used to measure colour is by direct comparison with a set of colour standards. This is attractive from the perspective that the human eye is very sensitive to deviations in colour. However, it can prove difficult to create colour standards which are not susceptible to change with age. Some 'liquid' standards have been suggested, including solutions of iodine and potassium dichromate as these have colour characteristics reminiscent of commercial beers. It is easily appreciated though that such standards are themselves not stable. For instance the orange dichromate is readily, if only slowly, reduced to green chromium(III).

Coloured glass discs provided a partial solution, with the first develop-

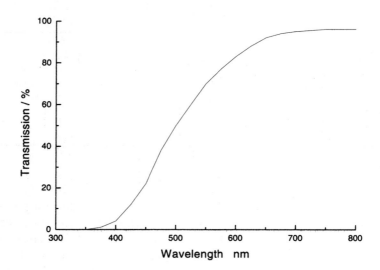

Figure 2.8   *Typical form of a transmission spectrum for an ale. Small differences (say a 5%
difference in the transmission spectrum over a small range of wavelengths)
results in a visibly distinct product*

ed by Lovibond in 1893. After a series of modifications, such comparator
discs were accepted by the European Brewery Convention in 1951. Dark
beers and products such as caramels can be readily assessed by making
the appropriate dilution, although the measurement was still subjective
as it required the human eye to compare discs and the beer sample. A
number of complications led to the proposal of various single
wavelengths for simple assessment of colour. Eventually the European
Brewery Convention settled on 430 nm, which was considered to be
appropriate for the paler type beers which dominated mainland Europe.
In the UK, 530 nm was chosen as this was around the wavelength where
maximum variation occurred between ales. A single figure though is
inherently limited as it cannot account for subtle variations in shade.

A development from a single wavelength determination is the calcula-
tion of tristimulus values.[16] Here, the primary red, green and blue colours
are represented by functions which express a relationship between
wavelength and intensity. These functions are designed to closely mimic
the response of the retinal cone cells. The red function shows a maximum
at around 600 nm, the green at 550 nm and blue at 450 nm and conform to
the internationally-recognised standards set by the Commission de Inter-
nationale l'Eclairage (CIE) and are known as the CIE colour-matching
functions. A transmission spectrum of the beer in a 10 mm cuvette relative

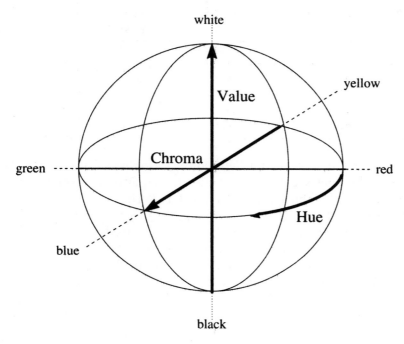

Figure 2.9 *The CIELAB colour space, showing the relationship between value, hue and chroma. Such a representation is possible because of the decomposition of the initial tristimulus values into three orthogonal axes*

to a water reference is used to derive each of the colour functions in turn. This on its own though is not sufficient to derive a standard colour, as colour perception depends on the source of illumination. There are a number of Standard Illuminants available. Of the two most common ones, Standard Illuminant C, represents typical daylight, and Standard Illuminant D_{65} includes extra energy in the UV region to derive more accurate measurements for fluorescent agents or optical brighteners. D_{65} is the most common Standard Illuminant used today for most colour determinations. After this processing, the result is three measurements representing red (X), green (Y) and blue (Z). These parameters are the tristimulus values.

Commonly these values are transformed into three further uncorrelated parameters L*, a* and b* which represent the colour in three-dimensional CIE L*a*b* (CIELAB) space (Figure 2.9). Hue is used to describe the predominant colour of an object, so that the hues red, yellow, green, blue and purple form a continuum in the equatorial plane of the colour space. Value is used to describe the brightness of a colour and is a vertical displacement between the two poles of the spherical colour space.

Table 2.2 *Representative CIELAB data for pairs of beers with similar EBC colour values. The human eye cannot detect* $\Delta E < 1$

Beer	L*	a*	b*	ΔE
A	95.00	−3.20	21.00	
B	94.50	−3.20	20.00	1.12
C	80.00	4.10	67.15	
D	77.50	5.15	68.00	2.84

From Figure 2.9, it is clear that light colours are displaced towards the white pole, whilst dark colours are below the equatorial plane towards the black pole. The third aspect of the CIELAB space is chroma – generally described as dull *vs* vivid or deep *vs* bright. Neutral or weak colours are found close to the North–South axis, whilst bright colours are located near the circumference of the colour space. Differences in colour of beers may be defined by the Euclidean distance between their positions in CIELAB space, or by ΔE (Table 2.2). Generally, as beer colour increases, it becomes more difficult to distinguish between such beers in terms of colour. For optimal (*i.e.* low colour, lager type beers) cases, $\Delta E > 1$ to enable reliable visual distinction to be made between beers.

The derivation of tristimulus values may appear to be somewhat numerically intensive, but it has been shown that, in practice, only five data points are required to interpolate the complete visible transmission spectrum of beer samples from which the L*a*b* parameters can be derived. The accuracy is sufficient for the measurement of beer colour.

BEER CLARITY

As has been mentioned previously, beer clarity is readily appraised by the consumer. The visual effects can vary, from a slight dulling of brightness to the observation of discrete particles. Here the discussion will be restricted to the factors which primarily affect the clarity of fresh beer, particularly the interplay between liquids and solids during the various brewing operations.

Brewing is by its very nature heterogeneous, frequently requiring the separation of solids and liquids in a controlled fashion. At the beginning of brewhouse operations, milled malt is blended with water, typically in a ratio of 1:3 (w/w) to effect enzymic activity and to extract the brew-positive components. Not all of the malt is solubilised, the residue being predominantly husk fragments. There is a need to substantially separate this insoluble material from the sweet wort, and this may be done either by mash filtration (if the grist is hammer milled) or by a lauter tun if the malt is milled more traditionally. For the latter, the husk fragments are allowed to accumulate on the coarse-slotted plate near the bottom of the lauter tun, effectively forming a filter bed. In each of these two cases, the solids are washed with hot water (typically 70 °C) to recover any residual extract. Over-exuberant application of water during this sparging operation can effectively extract significant amounts of polyphenols, so that there is a risk of conferring haze-potential on to the sweet wort.

This sweet wort is then boiled. The vigour of the boil is key for removing excess proteins and polyphenols (and some other minor components) as aggregates (trub or hot break). If the boil is not vigorous enough, again haze-potentiating material can remain in the beer. This hot wort is clarified by a whirlpool, where tangential entry of the hot wort into the tank creates a vortex in the centre of the accumulating liquid where the solid particles come together and sink to the bottom of the tank. The ideal degree of clarity of the exiting wort is a matter for debate. Very bright worts may be low in fatty acids and sterols, slowing down aerobic yeast growth during the first hours of fermentation.

To ferment the wort, solids are again added in the form of yeast cells. The ability of the yeast to flocculate simplifies the separation of the yeast from the immature beer. Again there is a compromise here: some yeast should remain in suspension for maturation, particularly for diacetyl reduction. Downstream adjustments of various parameters to bring the final beer into specification are often carried out before filtration – hop bitter acids, foam enhancers and colour products may be added. Thus, having processed the raw materials into unfiltered beer, there has at no stage been an absolute requirement for a bright liquid. In fact, solids are required at some stages. At filtration, solids are often introduced as a filter aid. On passage of the beer through the filter it should be within specification for haze. If this is not the case, remedial action may be necessary, as it is unlikely that the particulate material will spontaneously redissolve on storage, so the consumer will be presented with a hazy product. In short then, an effective filtration operation is crucial to the production of a bright product. Should components such as β-glucans remain in the beer until filtration, the filtration can suffer with a drop in filtration rate (flux) and short filter runs.

SUMMARY

For beer which is consumed from the glass, its foam is a key component on which the product is assessed. Its transient, colloidal composition has hampered detailed evaluation of beer foams. In addition, the components of beer foam, particularly the polypeptides and hop bitter acids have, until relatively recently, proved to be difficult analytes. Nevertheless, it is becoming increasingly apparent that amphiphilicity of both polypeptides and hop bitter acids is crucial for foam stability. Indeed, increasing the hydrophobicity of hop bitter acids by their chemical reduction to generate the tetra- and hexahydroiso-α-acids gives rise to improved foam stability.

The measurement of beer colour has evolved from a simple system for comparing a beer with a solid or liquid standard to a procedure today which takes into account not only the whole visible transmission spectrum of the beer, but also weights the data according to the sensitivity of the human eye. In the CIELAB system, the human eye can detect differences of $\Delta E > 1$ for pale beers, which is the Euclidean distance of two products in CIELAB space. Such an approach can be applied to bring beer colour reliably within specification.

The clarity of fresh beer is dependent on an effective filtration operation. The processes that go before can both help to protect the beer from haze formation during its shelf-life and enhance the filterability of the final beer.

REFERENCES

1 C.F. Bohren, in *Clouds in a Glass of Beer – Simple Experiments in Experimental Physics*, 1987, John Wiley & Sons, Inc., Chichester, UK.

2 P. Kaersgaard and J. Hejgaard, 'Antigenic beer macromolecules. An experimental survey of purification methods', *J. Inst. Brew.*, 1979, **85**, 103–111.

3 S. Yokoi, K. Maeda, R. Xiao, K. Kamada and M. Kanimura, 'Characterisation of beer proteins responsible for the foam of beer', *Proc. Eur. Brew. Conv. Cong.*, 1989, 593–600, IRL Press, Oxford.

4 D.C. Clark, P.J. Wilde and D. Marion, 'The protection of beer foam against lipid-induced destabilization', *J. Inst. Brew.*, 1994, **100**, 23–25.

5 S.B. Sorensen, L.M. Bech, M. Muldjberg, T. Beenfeldt and K. Breddam, 'Barley lipid transfer protein 1 is involved in beer foam formation', *Master Brew. Assoc. Am. Tech. Q.*, 1993, **30**, 136–145.

6 M. Meilgaard, 'Hop analysis, cohumulone factor and the bitterness of beer: Review and critical evaluation', *J. Inst. Brew.*, 1960, **66**, 35–50.

7 H. Lienert, *Proceedings of the European Brewery Convention Congress*, 1955, 282–289, IRL Press, Oxford.

8 G. Melm, P. Tung and A. Pringle, 'Mathematical modelling of beer foam', *Master Brew. Assoc. Am. Tech. Q.*, 1995, **32**, 6–10.

9 I.E. Crompton, P.K. Hegarty, *Proc. 3rd Aviemore Conf. on Malting, Brewing & Distilling*, ed. I. Campbell, Institute of Brewing, London,

1990, 277–281.

10 L.T. Lusk, H. Goldstein and D. Ryder, 'Independent role of beer proteins, melanoidins and polysaccharides in beer foam formation', *J. Am. Soc. Brew. Chem.*, 1995, **53**, 93–102.

11 C.W. Bamforth, 'The foaming properties of beer', *J. Inst. Brew.*, 1985, **91**, 370–383.

12 R.L. de Jong, 'A method to produce a beer foam with a defined bubble size distribution – heading for a new generation of foam analysis?', *Proceedings of the European Brewery Convention Congress*, 1995, 577–584.

13 D.R.J. Laws, J.D. McGuinness and H. Rennie, 'The loss of bitter substances during fermentation', *J. Inst. Brew.*, 1972, **78**, 314–321.

14 S.M. Smedley, 'Towards closer colour control in the brewery', *Brewers' Guardian*, 1995 (Oct.), 22–25.

15 S.M. Smedley, 'Towards closer colour control in the brewery. Part 2: A better approach to colour measurement', *Brewers' Guardian*, 1995 (Nov.), 44–47.

16 S.M. Smedley, 'Discrimination between beers with small colour differences using the CIELAB colour space', *J. Inst. Brew.*, 1995, **101**, 195–201.

Chapter 3

Flavour Determinants of Beer Quality

INTRODUCTION

For any food or beverage, the sensory characteristics are a vital parameter by which consumers appraise a product. This is important at the time of consumption, helping to shape consumer responses and preferences, and also when considering future purchases *i.e.* the extent of their loyalties to particular brands. The beer consumer has in fact tended to demonstrate significant brand loyalty (no doubt aided by effective marketing) but, nonetheless, sensory perception of beer is a complex, multidisciplinary issue which still requires substantial research input for a full understanding of the problem. In this chapter, the major flavour components typically found in beer will be described in terms of their sources and how these components classify in terms of taste, aroma and mouthfeel. Those compounds which arise primarily as a result of flavour instability during storage will be discussed in detail in Chapter 4. First though, the concept of the flavour unit, which will occur repeatedly throughout this chapter, needs to be defined.

The flavour unit (FU) is a useful, dimensionless number which provides an indication of the sensory level of a particular flavour attribute. It is calculated as shown in Equation 3.1.

$$\textit{Flavour unit (FU)} = \frac{[\textit{flavour compound}]}{\textit{sensory threshold}} \tag{3.1}$$

The exact value of the FU is dependent on how the sensory threshold is defined and measured and, particularly, if the threshold determination was carried out in the same or a similar product. If a compound is present at levels of less than one FU, then the compound of interest is present at levels below its sensory threshold. When a compound is present at 1–2

FUs, then it is detectable by the assessor. If a particular constituent is present at levels over two FUs, then it is likely to have a major effect on the sensory properties of the particular product being assessed. Throughout this chapter, typical levels of flavour compounds and their thresholds are given to help the reader to assess the importance of particular constituents to the overall flavour perception of beer. Much of this data has been assiduously compiled by Meilgaard.[1]

The final part of this chapter attempts to synthesise the diverse sources of beer flavour to aid in the definition of its major flavour impact components.

THE TASTE OF BEER

There are four tastes that are detected primarily by the tongue – sweet, salt, sour and bitter. There is also a fifth taste, called umami, described as a savoury taste and exemplified by monosodium glutamate. In addition to these, and of particular relevance to beer, is the stimulation of the trigeminal nerve which responds to irritants such as horseradish, onion, capsaicin and carbon dioxide. In fact, carbon dioxide levels in beer range from around $1\,g\,l^{-1}$ ($23\,mM$) for cask ales to more than $5\,g\,l^{-1}$ ($114\,mM$) for some lager products. Given that the sensory threshold of carbon dioxide in water is around $1\,g\,l^{-1}$, it is clear that carbon dioxide plays an important role in the sensory perception of beer. The evolution of carbon dioxide from beer can also affect the profile of volatile compounds which are present in the headspace above the beer. The use of nitrogen gas as a partial replacement for carbon dioxide in some stout and ale products gives rise to an appreciable softening of the flavour of the beer and can reduce the sensory impact of some flavour components present.

Sweetness

The sweetness of beer is a direct consequence of the presence of residual carbohydrates in the final product. These carbohydrates can arise either from the malt (if they have survived fermentation) or else they have been added as 'primings' to stimulate secondary fermentation to bring the beer up to specified carbon dioxide levels. The contribution of the most important carbohydrates to beer sweetness can be readily gauged (Table 3.1). If the carbohydrate consists of more than four glycosyl units it possesses little sweetness. However, these oligosaccharides can be beneficial to the perception of beer in that they contribute to body or mouthfeel by increasing the viscosity of the beer. The other consequence of having these more complex carbohydrates present is that they cannot be fer-

Table 3.1 *Sweetness and typical levels of some simple sugars found in beer.*
Sweetness is relative to that of sucrose (= 1.0)

Sugar	Relative sweetness	Range found in beers $(g\,l^{-1})$
Fructose	1.1	0–0.19
Glucose	0.7	0.04–1.1
Sucrose	1.0	0–3.3*
Maltose	0.5	0.7–3.0
Maltotriose	<0.5	0.4–3.4

*Found at higher levels in primed beers.

Table 3.2 *Representative organic acids found in beer*

Acid	Flavour threshold $(mg\,l^{-1})$	Typical levels in beer $(mg\,l^{-1})$	Flavour descriptors
Acetic	175	30–200	Acid, vinegar
Propanoic	150	1–5	Acid, vinegar, milk
Butanoic	2.2	0.5–1.5	Butter, cheese, sweat
2-Methylpropanoic	30	0.1–2	Sweat, bitter, sour
Pentanoic	8	0.03–0.1	Sweat, body odour
2-Methylbutanoic	2.0	0.1–0.5	Cheese, old hops, sweat
3-Methylbutanoic	1.5	0.1–2	—
Octanoic	15	2–12	Caprylic, goaty
Lactic	400	20–80	Acid
Pyruvic	300	15–150	Acid, salt, forage
Succinic	—	16–140	—

mented by brewers' yeasts. Indeed, it is common practice to use mixtures of simple and complex carbohydrates at mashing, as an adjunct to the malt, to control the fermentability of the resulting wort.

Sourness

The sourness of beer is, unsurprisingly, due to the presence of acids in the final beer. Typically beer pH ranges from 3.9–4.4, although some products do extend this range to some extent. The pH scale though is, of course, logarithmic, so that the range of proton activities which this scale represents are from around 40 up to 126 μM. It is also worth bearing in mind that strictly speaking the term pH is a misnomer for beer, as it is not generally a purely aqueous medium, typically containing 3 to 6% (v/v) ethanol (*i.e.* 0.5 to 1 M). This has a small but observable effect on the acid dissociation constants of weak organic acids, as might be expected because the dielectric constants of such aqueous ethanolic media will be less than that of pure water and therefore less able to maintain ions in solution. The major acids found in beer include carbonic acid, as well as acetic, lactic and succinic acids (Table 3.2), many of which have pK_a

Table 3.3 *Common inorganic ions and their effect on beer quality*

Ion	Typical levels in beer (mg l^{-1})	Sources	Effects
Potassium	200–450	Malt, adjuncts	Salt taste
Sodium	20–350	Brewing materials, water	Mawkish
Calcium	25–120	Brewing materials, water	Favourable effect on flavour
Magnesium	50–90	Brewing materials, water	Disagreeable with sulfate
Chloride	120–500	Water, starch hydrolysates	Full, sweet, mild flavour
Sulfate	100–430	Water	Dry
Oxalate	5–30	Mainly from malt	Haze, gushing inducer
Phosphate*	170–600	Malt, adjuncts	—
Nitrate	0.5–2.0	Water, hops	Off-flavour, high colour

*Most is taken up by yeast during fermentation.

values within two pH units of beer pH, so that small changes in pH can result in a large shift of the acid dissociation equilibrium. The final levels of these organic acids in beer, which come from the yeast, depend on the vigour of fermentation, with faster fermentations resulting in more of these acids being released. The sensory properties of such acids, though, are not restricted to proton activity. There are also solvated conjugate bases and undissociated acid molecules present which have sensory effects in addition to those of the protons released.

Saltiness

The salt taste of beer is a direct consequence of the presence of inorganic anions and cations (Table 3.3). Geographic regions historically famous for beer brewing, such as Burton-on-Trent, Pilsen and Munich were popular because the ionic content of the water which they used was particularly suited to the beers which they produced, even though the importance of the ionic content of water was not initially appreciated. There are also contributions of ions from the raw materials, particularly malt. Today, many brewers have a number of different sources of water for brewing and, to standardise water quality, often resort to deionisation (and, incidentally, dechlorination) of incoming water, before adding back the requisite ions. Particularly important are the concentrations of chloride and sulfate. Chloride provides a mellowing and fullness to the palate, while sulfate enhances the drying character of the beer.

Bitterness

One characteristic of beer recognised by virtually all consumers is its

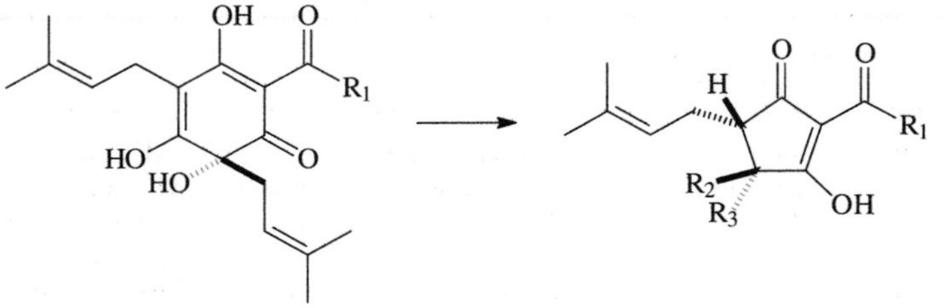

Figure 3.1 *The isomerisation of α-acids to iso-α-acids. The mechanism of the reaction is unclear, but the result is a ring contraction and the production of* cis- *and* trans-*isomers, with chiral induction at the prenyl branch point. R_1, R_2 and R_3 are defined in Table 3.4*

Table 3.4 *The major native iso-α-acids found in beer*

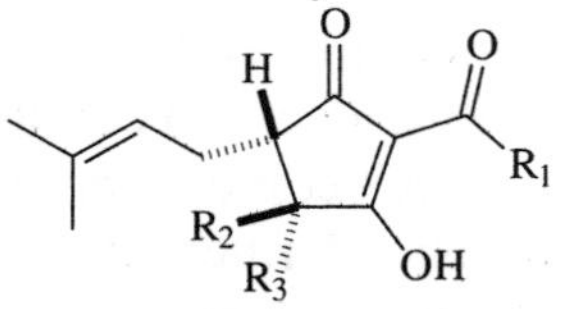

Compound	R_1	R_2^*	R_3^*	Typical proportions in beer (%)
trans-Isocohumulone	$CH(CH_3)_2$	OH	C_6H_9O	7
cis-Isocohumulone	$CH(CH_3)_2$	C_6H_9O	OH	30
trans-Isohumulone	$CH_2CH(CH_3)_2$	OH	C_6H_9O	10
cis-Isohumulone	$CH_2CH(CH_3)_2$	C_6H_9O	OH	40
trans-Isoadhumulone	$CH(CH_3)CH_2CH_3$	OH	C_6H_9O	3
cis-Isoadhumulone	$CH(CH_3)CH_2CH_3$	C_6H_9O	OH	10

*C_6H_9O is

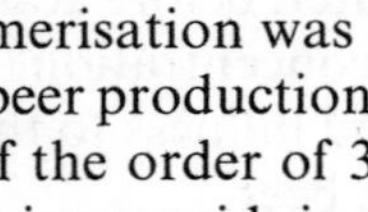

bitterness. This bitterness is largely attributable to a group of compounds called the iso-α-acids. These do not occur naturally in beer but are, in fact, isomerisation products of the naturally-occurring α-acids (Figure 3.1). This isomerisation was traditionally carried out during the wort boiling stage of beer production, but as the efficiency (utilisation) of isomerisation here is of the order of 30%, a processing industry has developed which produces iso-α-acids in virtually quantitative yields from the hop α-acids. These iso-α-acids are in fact a mixture of closely-related compounds, three pairs of *cis* and *trans* isomers, each pair deriving from a single α-acid (Table 3.4). The α-acid variants are thought to arise from the incorpor-

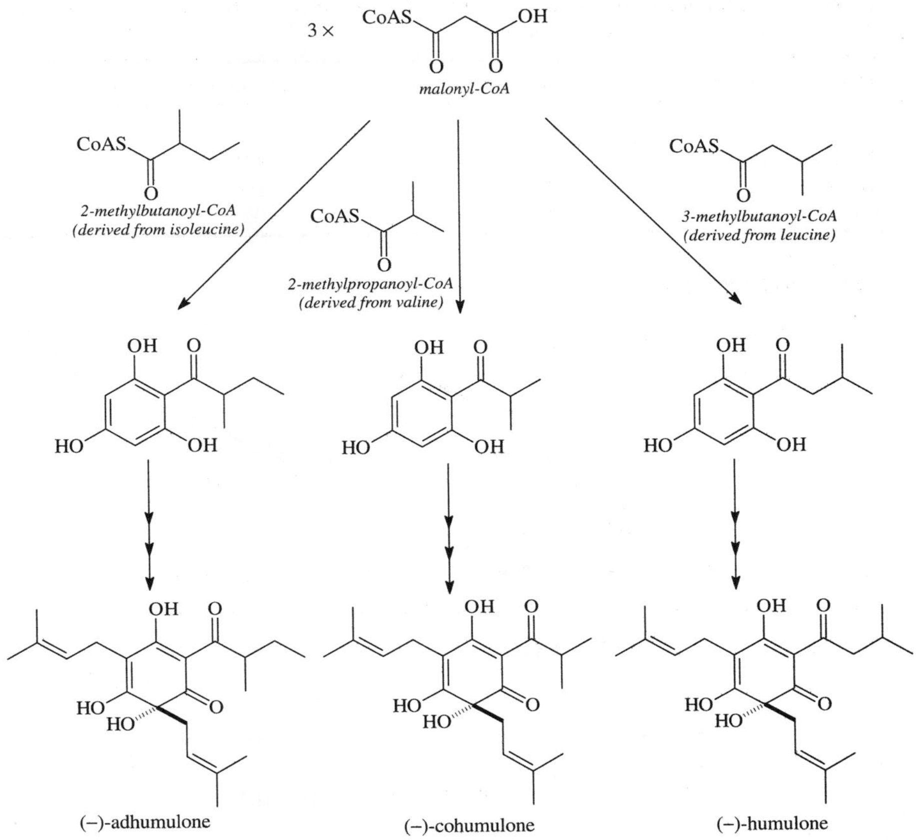

Figure 3.2 *Biosynthetic origin of the three major α-acids. The differences between these compounds is ultimately due to three different amino acid starting materials. Other minor α-acids can also be found in hops*

ation of either valine, leucine or isoleucine into the α-acid biosynthetic pathway (Figure 3.2). The mechanism of the ring contraction which occurs during isomerisation is unclear but the chiral tertiary carbinol centre appears to induce optical activity at the prenyl branch point. Partial racemisation at the migrating carbon results in the observed *cis* and *trans* pairs. The ratio of *cis/trans* isomers is dependent on the mode of isomerisation (Table 3.5), ranging from 80% *cis*-isomers for a magnesium catalysed isomerisation reaction, used by several processors, to 100% *trans*-isomers when isomerisation is effected with light (a process which is not carried out on a commercial scale).

Table 3.5 *Influence of the mode of α-acid isomerisation on the* cis/trans *ratio of the resultant iso-α-acids*

Means of isomerisation	cis-isomers (%)	trans-isomers (%)
Wort boiling	68	32
Aqueous alkali	55	45
Magnesium oxide	80	20
Light	0	100

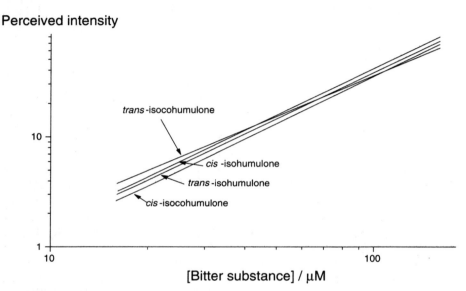

Figure 3.3   *Dose–response relationship between concentration and perceived bitterness of the* cis- *and* trans-*isomers of isocohumulone and isohumulone. The slopes of the lines are identical (p < 0.05) indicating that they are psychophysically equivalent. This may be interpreted as an indication that these compounds are detected on the tongue by the same mechanism. (NB The plot is in a log–log form as it is recognised that sensory dose–response relationships are non-linear.)*

There has been much debate about the importance of the detailed distribution of the various iso-α-acids and its ultimate effect on beer quality. Much of this has centred around the reported harsh bitterness which high proportions of the isocohumulones confer on beer flavour.[2] This is perhaps unexpected as bitterness intensity is considered to be related to hydrophobicity, and the isocohumulones are the most polar of the iso-α-acids. More recently, sensory studies with the purified compounds have shown that the psychophysical behaviour of the individual compounds is similar (Figure 3.3) but that the isocohumulones are actually significantly less bitter than their more hydrophobic counterparts. Time–intensity studies of the purified compounds also showed that there

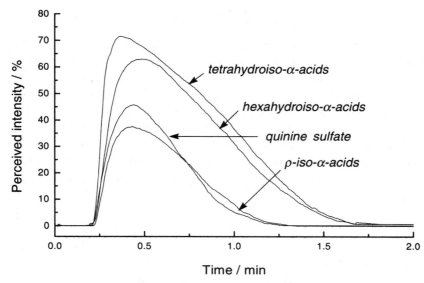

Figure 3.4 *Time–intensity curves for the chemically-modified iso-α-acids and quinine sulfate. These commercial samples were tested at 33 μM in aqueous buffer at pH 4.2, whilst the quinine was tested at 26 μM in the same medium*

was little evidence to suggest that the isocohumulones lingered on the palate any longer than the other iso-α-acids. Furthermore, chemically-modified iso-α-acids are now commercially available for bittering beer which have different hydrophobicities relative to the native compounds. These more clearly demonstrate the positive correlation between bitterness and hydrophobicity (Figure 3.4). Intriguingly, whilst there is little difference in the hydrophobicities of *cis* and *trans* iso-α-acid pairs, the *cis*-isomers appear to be significantly more bitter than their *trans*-counterparts.[3] Typical levels of native iso-α-acids in beers are $10–60 \, mg \, l^{-1}$, although extremes do exist outside these ranges. Given that their sensory detection threshold is around $5–6 \, mg \, l^{-1}$, it is apparent that they are important to beer flavour, being present at levels generally greater than two flavour units.

The chemistry of the iso-α-acids and their precursors is extremely complex, but the degradation of the flavour-neutral β-acids gives low yields of the hulupones (up to $6 \, mg \, l^{-1}$ in some beers), which are reported to have bitterness values of similar intensity to iso-α-acids. There are also measurable quantities of the *allo*-iso-α-acids, which have been reported to be present at levels greater than one flavour unit. As yet though, these relatively minor bittering compounds have not been subject to the same scrutiny as the native iso-α-acids.

BEER AROMA

Components in beer often considered to be tastes are in fact detected by the nose. The misconception arises because of the connectivity of the tongue, throat and the nasal passages. Beer aroma therefore arises not only from sniffing it, but also by the distillation of volatiles from the beer when taken into the mouth. Because of the chemical complexity of beer, it is perhaps unsurprising that beer aroma is not characterised by one or even a few well-defined components. Instead, many compounds contribute to beer aroma, both individually and in a synergistic or antagonistic sense. Below, the various classes of components and their sources are discussed in turn.

Esters

There are a number of esters which contribute to the flavour of beers (Table 3.6). Of these components, those of most concern to the brewer are ethyl acetate and the so-called iso-amyl acetate, which is in fact a mixture of 2- and 3-methylbutyl acetates. The levels of these compounds in beer are influenced by a number of factors, including the gravity (density) of the wort and the amount of oxygen to which the yeast is exposed. This is because the esters are formed from their respective alcohols under conditions when the acetate group is available but is not required for the biosynthesis of key components (*i.e.* lipids) of the yeast membranes. Thus, factors which promote yeast production, such as high oxygen levels and low wort lipids, lower ester production and *vice versa*. However, perhaps the major factor which affects the extent to which esters are produced is the yeast strain itself, with some strains more readily generating esters than others.

Alcohols

A range of alcohols affect the flavour of beer (Table 3.7). The most important of these is ethanol, which is present in most beers at levels at least two orders of magnitude higher than any other alcohol. Ethanol contributes directly to the flavour of beer, giving rise to a warming character. Another, more obvious descriptor, is alcoholic! Ethanol also plays a role in the flavour perception of other beer components. It can influence the partitioning of flavour components between the liquid beer, foam and the headspace above the beer. Thus the production of low- or non-alcoholic beers is not simply a matter of removal, or prevention of

Table 3.6 *Significant flavour-active esters found in beer*

Ester	Typical levels $(mg\,l^{-1})$	Flavour threshold $(mg\,l^{-1})$	Flavour descriptors
Ethyl acetate	10–60	30	Solvent-like, sweet
Isoamyl acetate	0.5–5.0	1	Banana, ester, solvent
Ethyl hexanoate	0.1–0.5	0.2	Apple, fruity, sweet
Ethyl octanoate	0.1–1.5	0.5	Apple, tropical fruit, sweet
2-Phenylethyl acetate	0.05–2.0	3.0	Roses, honey, apple, sweet
Ethyl nicotinate	1.0–1.5	2 (?)	Grainy, perfume

Table 3.7 *Selected alcohols commonly found in beers*

Alcohol	Typical levels $(mg\,l^{-1})$	Flavour threshold $(mg\,l^{-1})$	Flavour descriptors
Methanol	0.5–3.0	10 000	Alcoholic, solvent
Ethanol	20 000–80 000	14 000	Alcoholic, strong
1-Propanol	3–16	700	Alcoholic
2-Propanol	3–6	1500	Alcoholic
2-Methylbutanol	8–30	65	Alcoholic, vinous, banana
3-Methylbutanol	30–70	70	Alcoholic, vinous, banana
2-Phenylethanol	8–35	125	Roses, bitter, perfumed
1-Octen-3-ol	0.03	0.2	Fresh-cut grass, perfume
2-Decanol	0.005	0.015	Coconut, aniseed
Glycerol	1200–2000	—	Sweetish, viscous
Tyrosol	3–40	200	Bitter, chemical

the formation of, significant amounts of ethanol, but rather requires some form of (as yet poorly defined) modification to adjust for the lack of ethanol.

The higher alcohols (*i.e.* of higher molecular weight than ethanol) are important as the immediate precursors of the more flavour-active esters, so that the control of higher alcohol formation needs regulation to ensure that, in turn, ester production is controlled. The higher alcohols are produced by yeast as secondary metabolites of amino acid metabolism (Figure 3.5). Unsurprisingly then, the levels of free amino nitrogen (FAN) in the wort influence the levels of higher alcohols formed. The situation is actually complicated by the fact that yeast cells are capable of synthesising their own higher alcohols from other pathways rather than from amino acids. Suffice to say that higher alcohol production is increased at both excessively high and insufficiently low levels of assimilable nitrogen available to the yeast from the wort. Again, as for esters, yeast strain turns out to be the most important factor, with ale strains producing more

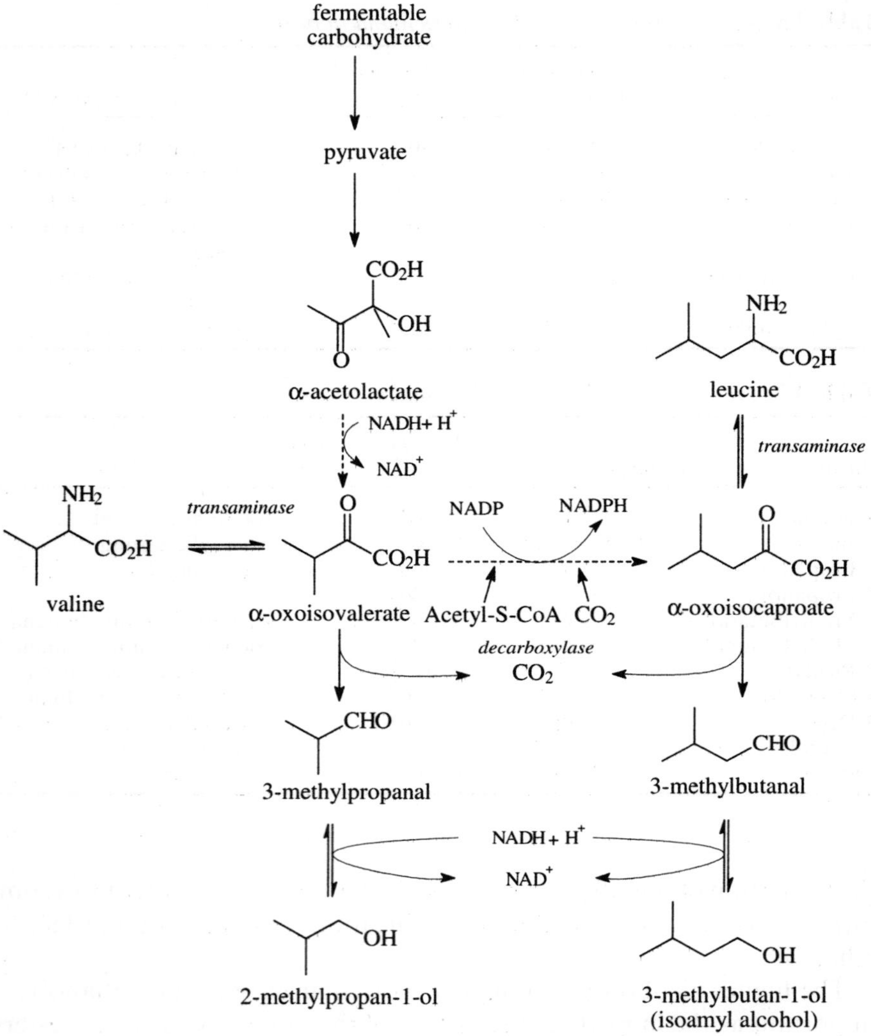

Figure 3.5 *The formation of 2-methylpropan-1-ol and 3-methylbutan-1-ol from carbohydrate metabolism and the biosynthesis of valine and leucine*

higher alcohols than lager strains. Conditions which favour increased yeast growth, such as excessive aeration or oxygenation, promote higher alcohol formation, but this can be ameliorated by the application of a top pressure during fermentation.

Table 3.8 *Vicinal diketones and their reduced components found in beer*

Vicinal diketone	Typical levels $(mg\,l^{-1})$	Flavour threshold $(mg\,l^{-1})$	Flavour descriptors
2,3-Butanedione	0.01–0.4	0.07–0.15	Butterscotch
3-Hydroxy-2-butanone	1–10	17	Fruity, mouldy, woody
2,3-Butanediol	50–150	4500	Rubber, sweet, warming
2,3-Pentanedione	0.01–0.15	0.9	Butterscotch, fruity
3-Hydroxy-2-pentanone	0.05–0.07	—	—

Vicinal Diketones

Whilst the esters and higher alcohols in beer can often be regarded as providing a positive contribution to beer flavour the same cannot be said of the so-called vicinal diketones (VDKs).[a] One exception is the acceptance of these compounds in some UK ales and some niche products. The major VDK in beer is 2,3-butanedione (diacetyl), but there are also significant quantities of 2,3-pentanedione produced during fermentation (Table 3.8). Elimination of VDKs from beer is dependent on fermentation management. These substances ultimately derive from pyruvate and its homologue α-oxobutyrate (Figure 3.6). The precursor α-acetolactate is thought to leak out of the yeast and break down spontaneously to form VDKs. Nevertheless, yeast cells can take up VDKs from their surrounding medium, providing of course that the yeast remains in contact with the beer and that the yeast is healthy. Diacetyl is reduced sequentially first to acetoin and then to butane-2,3-diol, with pentane-2,3-dione undergoing analogous reduction reactions. Both acetoin and butane-2,3-diol are much less flavour-active than their VDK precursor. Allowing the temperature to rise at the end of fermentation facilitates the rapid removal of VDKs. Alternatively, in a process known as Krausening, a small amount of freshly fermenting wort is added late on as an inoculum of healthy yeast. Another approach is that taken by the Finnish brewer Sinebrychoff, where the acetolactate precursors are decomposed by heating beer in the absence of oxygen[b] prior to the use of an immobilised yeast to consume the free VDKs. Yet another route is to use the enzyme, acetolactate decarboxylase, which can be added during fermentation, to convert acetolactate directly to acetoin avoiding the intermediate and much more

---

[a] Although nowadays termed α-diketones, the term vicinal diketones persists in the brewing industry and will be used here.
[b] The minimisation of oxygen is essential if colour and flavour changes are to be minimised.

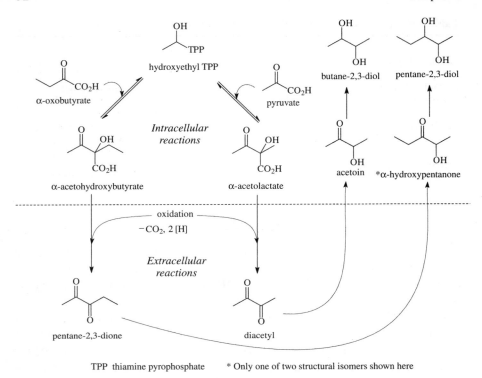

Figure 3.6 contents labels:

OH
TPP
hydroxyethyl TPP

O
CO₂H
α-oxobutyrate

O
CO₂H
pyruvate

OH

OH
butane-2,3-diol

OH

OH
pentane-2,3-diol

Intracellular reactions

O
OH
CO₂H
α-acetohydroxybutyrate

O
OH
CO₂H
α-acetolactate

O

OH
acetoin

O

OH
*α-hydroxypentanone

— oxidation —
−CO₂, 2 [H]

Extracellular reactions

O

O
pentane-2,3-dione

O

O
diacetyl

TPP thiamine pyrophosphate   * Only one of two structural isomers shown here

Figure 3.6 *Pathways for the formation and subsequent breakdown of vicinal diketones (VDKs) in beer*

flavour-active diacetyl. If brewers experience a persistent difficulty with diacetyl in the final product, it may be indicative of an infection either by *Pediococcus* or *Lactobacillus* bacteria.

Sulfur Compounds

Some of the most characteristic flavours of beer are due to the presence of sulfur compounds. There are many of these (Table 3.9) and they make various contributions to beer flavour. Many ales contain appreciable amounts of hydrogen sulfide. To some, this may seem a rather unpleasant thought, but at levels a little above the sensory threshold, its flavour impact is not unpleasant and is in fact characteristic of ales. Of course, high levels of hydrogen sulfide are to be avoided, because of the resulting rotten eggs or drains attributes which it imparts to beer flavour! Hydrogen sulfide develops in beer by a number of routes (Figure 3.7). These include the reduction of sulfite or sulfate, the breakdown of amino acids (*e.g.* cysteine) or peptides such as glutathione. The reductive activities are

Table 3.9 *Volatile sulfur compounds commonly found in beer*

Sulfur compound	Typical levels ($\mu g\,l^{-1}$)	Flavour threshold ($\mu g\,l^{-1}$)	Flavour descriptors
Hydrogen sulfide	1–20*	5	Sulfidic, rotten eggs
Sulfur dioxide	200–20 000*	> 25 000	Sulfitic, burnt match
Carbon disulfide	0.01–0.3	> 50	—
Methanethiol	0.2–15	2.0	Putrefaction, drains
Ethylene sulfide	0.3–2.0	> 20	—
Ethanethiol	0–20	1.7	Putrefaction
Propanethiol	0.1–0.2	0.15	Putrefaction, rubber
Dimethyl sulfide	10–100	30	Sweetcorn, tin tomatoes
Diethyl sulfide	0.1–1.0	1.2	Cooked vegetables
Dimethyl disulfide	0.1–3	7.5	Rotten vegetables
Diethyl disulfide	0–0.01	0.4	Garlic, burnt rubber
Dimethyl trisulfide	0.01–0.8	0.1	Rotten vegetables, onion
Methyl thioacetate	5–20	50	Cabbage
Ethyl thioacetate	0–2	10	Cabbage
Methionol (1; 4-thia-1-pentanol)	50–1300	2000	Raw potatoes
Methional (2; 4-thiapentanal)	20–50	250	Mash potatoes, soup-like
3-Methyl-2-butene-1-thiol	0.001–0.1	0.01	Skunk, leek-like, lightstruck

*Total levels (free and bound) quoted.

(1)   (2)

again dependent on yeast strain. Thus ale strains are generally more adept at reducing sulfur dioxide to hydrogen sulfide than their lager counterparts. However, the volatility of hydrogen sulfide (boiling point = −60 °C) means that control of the process, particularly the rate of carbon dioxide evolution, can help lower hydrogen sulfide to reasonable levels. This does of course require a healthy yeast and a vigorous fermentation. Hydrogen sulfide can also arise as a by-product of yeast autolysis, which will be more prevalent for unhealthy yeast. Cask ales often contain appreciable levels of hydrogen sulfide. This can be due to the action of healthy yeast on the sulfite which is present as a stabiliser in many of the 'finings' used to clarify such products. Hydrogen sulfide in beer can also be due to microbial infection. For instance, *Zymomonas*

sulfur oxidation state

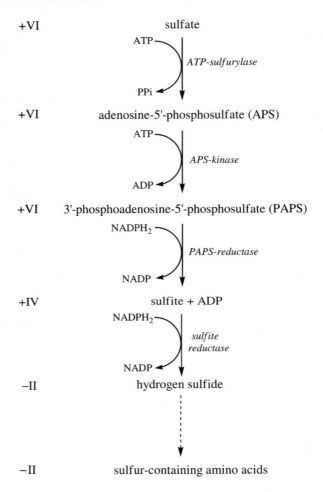

+VI	sulfate
+VI	adenosine-5'-phosphosulfate (APS)
+VI	3'-phosphoadenosine-5'-phosphosulfate (PAPS)
+IV	sulfite + ADP
−II	hydrogen sulfide
−II	sulfur-containing amino acids

Figure 3.7 *Simplified pathway for the formation of hydrogen sulfide in beer. Other pathways, such as enzymatic demethylation of DMS and methanethiol are also likely to be a source of hydrogen sulfide. ATP = adenosine triphosphate, ADP = adenosine diphosphate, NADP = nicotinamide adenine dinucleotide phosphate, PPi = inorganic phosphate)*

produces not only hydrogen sulfide, but also acetaldehyde.

The sulfur character of lager beers tends to be more complex. The major sulfur flavour is usually dimethyl sulfide (DMS). Most DMS in beer arises from the degradation of malt-borne *S*-methylmethionine (SMM) which in turn is generated during the proteolysis which occurs at malting. SMM is thermally labile and readily breaks down to give DMS, so that SMM survives to a greater extent in mildly kilned malt. Therefore, the ultimate levels of DMS tend to be higher in lagers than ales, as malt for the latter is kilned at slightly higher temperatures to generate more colour. SMM is extracted from the malt grist during mashing and its survival throughout the process is dependent on processing. A vigorous boil will cause more SMM to be converted to DMS and this DMS will be lost by evaporation. During whirlpool operations, conditions of temperature and agitation (shear) are less vigorous so that any DMS produced here is more likely to be retained in solution. It is not uncommon for brewers to specify the level of SMM in their malt so that they can more readily achieve final DMS specifications. Process operations, particularly the boil and whirlpool stages, can then be manipulated to achieve specified DMS levels for the pitching wort. It is important to appreciate that the specified wort DMS levels will be somewhat higher than those for beer, as the volatility of DMS results in significant losses during fermentation as a result of its being purged during the evolution of carbon dioxide.

There are two further points that should be borne in mind when considering DMS. Firstly, some SMM is converted to dimethyl sulfoxide (DMSO) during malt kilning. This is neither heat labile or particularly volatile, but it is water-soluble. It can be transferred into wort at appreciable levels and some yeast strains can reduce DMSO to DMS. Secondly, late hopping with pellets or whole hops can contribute up to $15 \, \mu g \, l^{-1}$ of DMS to beer.[4] Thus the control of DMS in the final product requires a holistic view of beer production – encompassing process, raw materials and yeast – to ensure that the specified levels of DMS find their way into the final product. A twist on the DMS theme was the observation that the presence of 2-phenylethanol or 2-phenylethyl acetate suppressed the flavour intensity of the DMS present in a dose-dependent manner.[5] There is no apparent explanation for this, but this observation does highlight the complexity of sensory perception and the need for sensory evaluation as well as analytical information to ensure that a more complete picture of beer sensory qualities is drawn.

Another important sulfur compound, which is derived from the bitter iso-α-acids, is a highly flavour-active allylic thiol, 3-methyl-2-butene-1-thiol (MBT). This compound is formed by the photochemical degradation of iso-α-acids in the presence of a photosensitiser (often cited as riboflavin) when a source of sulfur is present. The most critical property of MBT is its flavour threshold, which is typically $10 \, ng \, l^{-1}$. The sheer flavour intensity of this compound poses significant challenges to the analyst who wishes to study its formation and behaviour, which perhaps accounts for the uncertainty surrounding the mechanism of its formation (Figure 3.8). MBT develops rapidly in beer exposed to light, and can

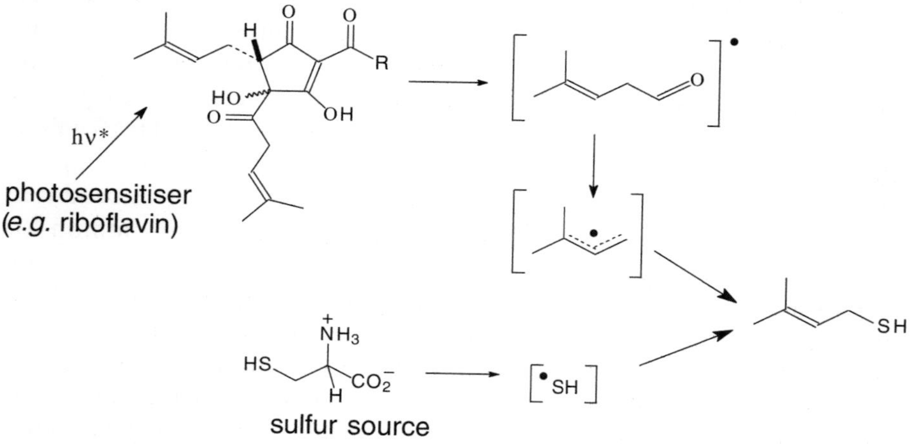

Figure 3.8 *Postulated mechanism for the formation of 3-methyl-2-butene-1-thiol (MBT) in beer*

readily attain sensory levels of greater than 10 flavour units. It is de-scribed variously as a rather uncomplimentary skunky aroma and also leek-like. Its development at such high levels (in sensory terms) means that, for affected products, this flavour attribute can dominate above all others. Protection of beer from the formation of MBT is effected in one of two ways. Firstly, physical barriers to light can be used, such as brown (but not clear or green) glass, cardboard wraps for bottle packs, large labels and, of course, containment in opaque packages, which are all effective solutions. An alternative is to reduce the reactivity of the iso-α-acids towards photochemical degradation. It has become apparent that, for MBT to form, both the carbonyl and unsaturated moieties of the isohexenoyl side-chain are essential. These can be removed by chemical modification. The aliphatic carbon–carbon double bonds are susceptible to palladium-catalysed hydrogenation, which gives the substantially more bitter and hydrophobic tetrahydroiso-α-acids (Figure 3.9(a)). Alter-natively, the carbonyl group can be reduced to the corresponding alcohol with sodium borohydride to give the ρ-iso-α-acids (Figure 3.9(b)) and, perhaps unsurprisingly, both treatments, yielding the hexahydroiso-α-acids (Figure 3.9(c)), are also effective. In fact all of these materials are available commercially for addition to beer. However, for light stability, 100% replacement of the native iso-α-acids is required as well as the rigorous exclusion of these latter from the brewing plant. Whilst this may appear straight-forward, the entrainment of iso-α-acids in beer foam can

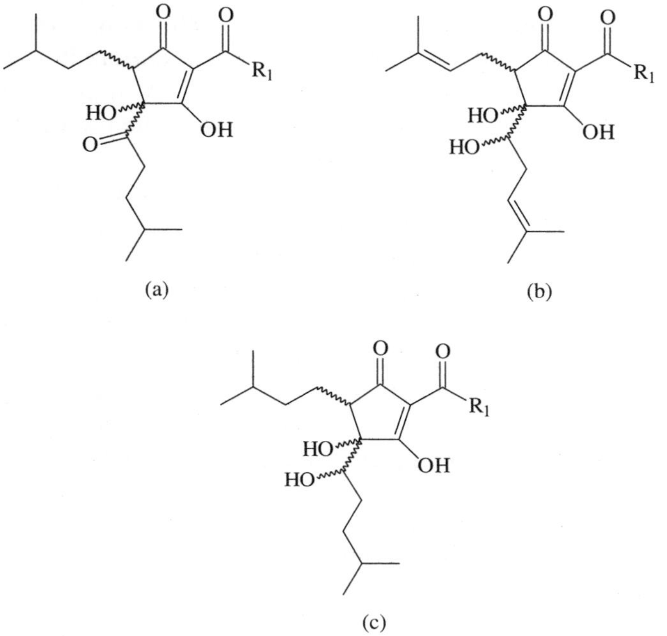

Figure 3.9 *Structures of the chemically-modified iso-α-acids used for partial or total replacement of the native iso-α-acids traditionally used for the production of beer. Variations in R_1 are equivalent to those for the iso-α-acids (R_1) in Table 3.4. (a) Tetrahydroiso-α-acids, (b) ρ-iso-α-acids, (c) hexahydroiso-α-acids*

result in brewing vessels and pipework having small, but significant quantities of incompletely removed foam adhering to their walls, providing a ready source of light instability. Another source of 'contamination' is the use of yeast propagated conventionally, *i.e.* in wort containing native iso-α-acids. Whilst the flavour activity of MBT means that only one iso-α-acid molecule per million needs to be converted to MBT to cause a noticeable flavour defect, in fact residual iso-α-acid levels of $10–100\,\mu g\,l^{-1}$ are generally considered to confer adequate light stability on the final beer.

Whilst the presence of MBT at elevated levels in beer is no doubt apparent to the consumer, its presence is detectable in almost all beers analysed by gas chromatography coupled to an olfactory port. This perhaps indicates that MBT is a significant odorant of beer and helps to make beer recognisable as such to the consumer. Indeed, such a claim has been made for coffee, where MBT has been found to be a key flavour attribute for authentic coffee aroma.[6]

The presence of sulfur dioxide in beer is considered to enhance the keeping qualities of beer (see Chapter 4). Its sensory properties are quite different from those of the other volatile sulfur compounds, not only in terms of is sensory description (sulfitic, burnt match) but also its flavour threshold. Nevertheless, suprathreshold levels of sulfur dioxide can exist in beer and has been a cause for concern with the increased incidence of allergic reactions by sensitive individuals.

Other sulfur compounds are also present in beer (Table 3.9). These fall into several categories: thioesters, sulfides (such as DMS), disulfides, polysulfides, thiols and sulfur-containing heterocycles. A range of simple thiols are found in beer, particularly methanethiol, ethanethiol and pro-panethiol. They all generally have low flavour thresholds and unpleasant sensory characteristics. Methanethiol is thought to combine with active acetate to produce methyl thioacetate. The thioesters are generally less flavour-active and have less offensive flavour attributes, such as cooked vegetable. Sulfur-bearing heterocycles are presumably Maillard reaction products and can be formed either during malt kilning or wort boiling. The approaches for the control of these flavour compounds are by no means clear as yet and it is likely that future research activity in this area will be highly beneficial to assist the brewer to manipulate and control beer flavour.

Hop Aroma

The addition of hops to beer not only imparts bitterness but also a unique 'hoppy' aroma. There has been some debate as to whether this arises from the breakdown of the non-volatile hop components, such the α- and β-acids, or from the oil constituents. It is now generally accepted that the latter is the predominant source of the hoppy aroma in beers. The difficulties in understanding the provenance of hoppy flavours in beer is due to the complexity of hop oil, which contains in excess of 300 compo-nents[4] (Table 3.10). Before considering what constitutes hoppy aroma in beer, it is worth reviewing the composition of hop oil itself.

Of this oil, around 40–80% is hydrocarbon, which, with the exception of some simple aliphatic hydrocarbons, is mainly terpene in origin (Figure 3.10). Monoterpenes are represented by acyclic, monocyclic and bicyclic compounds, but myrcene is the most abundant and can account for up to 30% of the total oil. There is a group of at least 40 sesquiter-penes, which again can be either acyclic, mono-, bi- or tricyclic. The major components from this group are usually humulene and caryophyllene. Only two diterpenes have been identified in hop oil – *m*- and *p*-camphor-ene – both of which are Diels-Alder adducts formed from two molecules of myrcene.

Table 3.10 *Classification of hop oil components (after Moir[4])*

Classification	Approximate number identified
Hydrocarbons	60
Aldehydes	20
Ketones	50
Esters	70
Acids	10
Alcohols	70
Oxygen heterocycles	30
Sulfur compounds	30

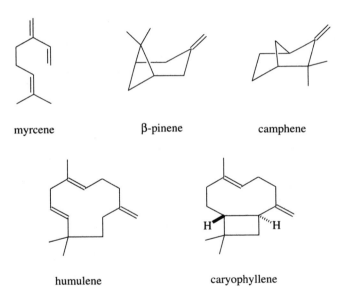

myrcene   β-pinene   camphene

humulene   caryophyllene

Figure 3.10 *Representative terpenes found in hop oils*

Some saturated and unsaturated aldehydes have been reported as hop oil constituents, as well as the monoterpenoid citral a and b (geranial and neral respectively) and citronellal (Figure 3.11). Aldehydes, particularly when unsaturated, are often highly flavour-active (see Chapter 4), but in fact are reduced to their corresponding alcohols during fermentation. Indeed, the presence of yeast aldo- and ketoreductase activities during fermentation has been demonstrated.[7]

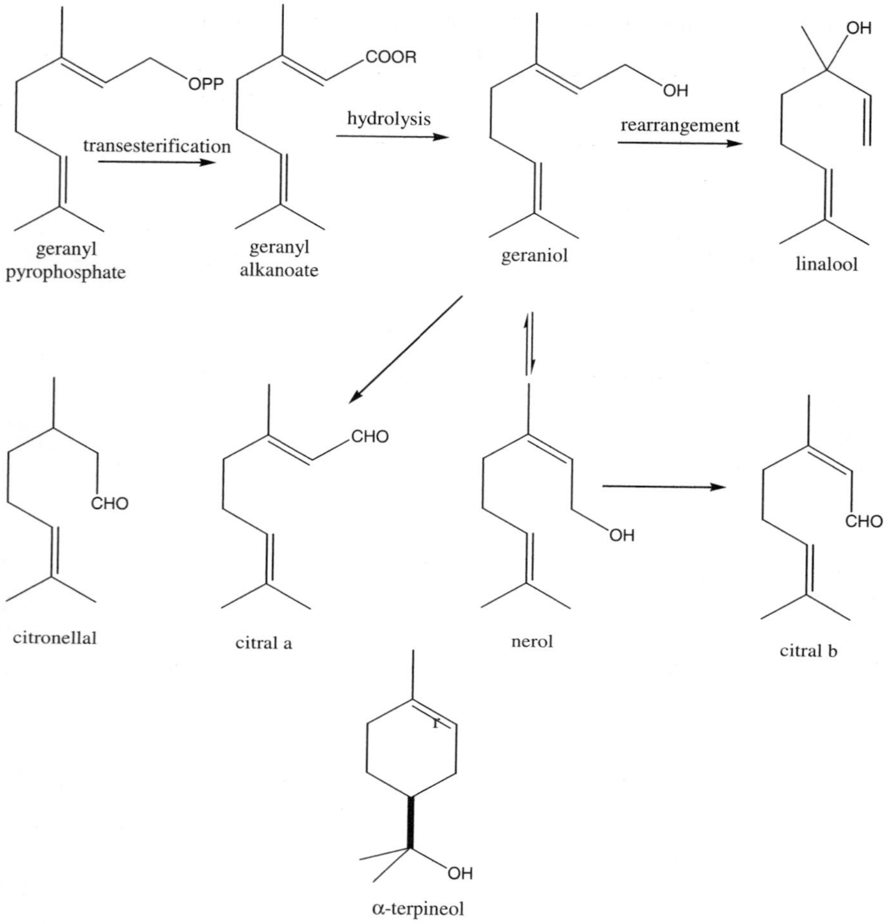

Figure 3.11 *Relationships between monoterpenoid compounds found in hop oils*

There are a large number of methyl ketones which have been reported to occur in hop oil. A homologous series from heptan-2-one to heptadeca-2-one is evident in most hop varieties, together with a number of branched and unsaturated ketones. There is also some evidence for the presence of alkane-2,4-diones in the hop oil of one variety, Wye Target. The highly flavour-active nor-carotenoids, β-ionone and β-damascenone, have been identified in hop oil and found in beer at levels which indicate that they may be important contributors to hop flavour in beer.

The esters identified in hop oil include a homologous series of methyl esters, from hexanoate to dodecanoate, as well as a number of branched-chain and unsaturated methyl esters. Most of the hop oil esters are hydrolysed during fermentation, and transesterification in the presence of increasing levels of ethanol has been shown to occur. Methyl esters of some conjugated acids, such as methyl geranate, resist hydrolysis and are therefore detectable in beer.

Some simple aliphatic carboxylic acids are sometimes present in trace quantities in hop oil, but the presence of branched-chain acids, (such as 2-methylbutanoic acid and 4-methyl-3-pentenoic acid), has been shown to be due to the oxidation of the α- and β-acids.

There are a wide range of alcohols present in hop oil, from series of straight- and branched-chain alkan-1-ols, to terpene alcohols. Linalool is the major terpene alcohol found in hops and can account for up to 1% of the total oil. Isomeric compounds include geraniol (largely esterified in hop oil) and α-terpineol (Figure 3.11). Moir[4] suggested that a useful distinction could be made between terpene alcohols which can be considered to be the end-products of biosynthetic pathways (*e.g.* linalool) and allylic terpene alcohols (such as β-humulene-1-ol) which are terpene oxidation products and can be prepared by rearrangement of their epoxide counterparts. Levels of the former tend to fall during hop storage while levels of the latter increase.

The chemical lability of the sesquiterpene hydrocarbons is evident from their behaviour during hop storage. Thus the levels of these hydrocarbons decrease on storage, whilst levels of epoxide auto-oxidation products increase. Humulene-4,5-epoxide has been identified in a commercial beer, and a number of humulene diepoxides have been detected in both hops and beer (Figure 3.12).

Some cyclic ethers have been detected in hop oil, including the *cis*- and *trans*-linalool oxides. These are presumably formed by the cyclisation of linalool-6,7-epoxide (Figure 3.13). Such components are efficiently transferred to wort during the wort boiling process and, together with other compounds such as karahana ether and hop ether, are considered to be important hop flavour compounds. The 3(2*H*)-furanones, found in both hop oil and beer, show structural homology which indicates that they are derived by the degradation of the α- or β-acids (Figure 3.14).

Hops also contain a range of sulfur compounds. Green hops contain two series of *S*-alkyl thioesters, the only volatile sulfur compounds present. Their levels are essentially unchanged during kilning and resist both hydrolysis and transesterification. There are a number of literature reports of various sulfur compounds found in hops which are artefacts formed by the reaction of terpene hydrocarbons with residual sulfur which has been used as a fungicide to protect against powdery mildew damage. This usually does not result in any flavour problems, but a more

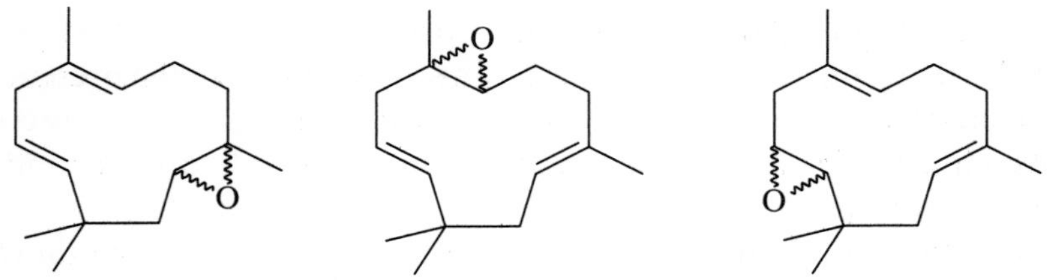

humulene epoxide I
(humulene-8,9-epoxide)

humulene epoxide II
(humulene-1,2-epoxide)

humulene epoxide III
(humulene-4,5-epoxide)

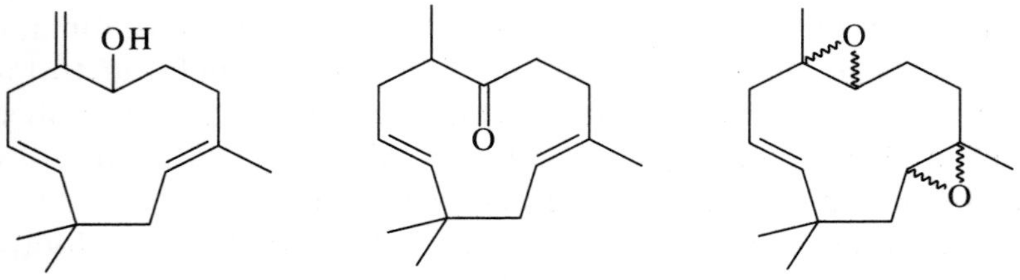

humulenol II
(β-humulen-1-ol)

humuladienone

humulene diepoxide III
(humulene-1,2,4,5-diepoxide)

Figure 3.12 *Oxidation products typical of those found in hop oils and beers*

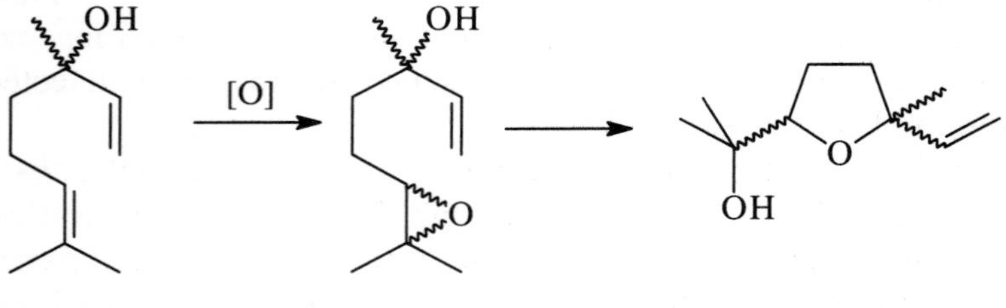

(±)-linalool

linalool-6,7-epoxide

cis- and *trans*-linalool oxides

Figure 3.13 *Formation of linalool oxides from linalool*

serious issue of the presence of free sulfur on hops is the ability of yeast to generate sulfury and burnt off-flavours during fermentation when the hops are used for late or dry hopping.

From what has been said about the complexity, volatility and chemical lability of various groups of hop oil components, it is perhaps unsurprising that hoppy aroma in beer is far from understood. Indeed, it is likely that compounds which are considered to be hop-derived that occur in beer do not actually occur in hops themselves. A number of hop-derived

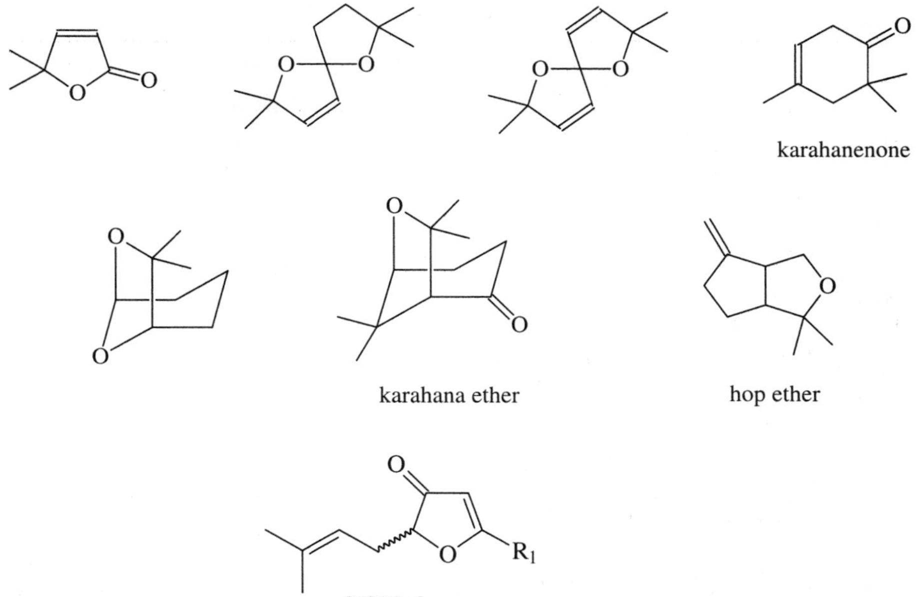

karahanenone

karahana ether

hop ether

3(2*H*)-furanones
(R₁ as defined in Table 3.4)

Figure 3.14 *Miscellaneous oxygen heterocycles found in hop oils*

compounds have been measured in beers, together with their thresholds (Table 3.11). There is a need though to treat such information with caution, as the analytical values are heavily dependent on the beer type and processing conditions employed, and the way in which the analysis was carried out. Furthermore, threshold values tend to be product specific and rely on the purity of the compounds which have been assessed.

The exact make-up of hoppy flavour in beer depends on where the hops or hop products are employed. Thus, to impart bitterness, it is usual to add hops, or more commonly hop pellets, at the beginning of the boil. The hoppy aroma which results from this is referred to as kettle hop aroma. Some workers have reported the loss of 95% of the total hop oil within five minutes of addition of the hops to a boil. Thus it is clear that few hop oil components survive the boil, although other workers have considered the hop acid degradation product 2-methyl-3-butene-2-ol to be a kettle hop character. In addition to this is a vast array of oxygenated sesquiterpenes, virtually all of which are below their flavour thresholds. For the production of lager beers, it is common practice to add a portion of the hops (typically 20%) towards the end of the boil. While this does not allow sufficient time for the effective conversion of the α-acids to their

Table 3.11 *Some hop oil-derived compounds reported in beers*

Compound	Range of reported levels ($\mu g\, l^{-1}$)*	Flavour threshold in beer ($\mu g\, l^{-1}$)
Linalool	1–470	27, 80
Linalool oxides	nd–49	—
Citronellol	1–90	—
Geraniol	1–90	36
Geranyl acetate	35	—
α-Terpineol	1–75	2000
Humulene epoxide I	nd–125	10#
Humulene epoxide II	1.9–270	450
α-Eudesmol	1–100	—
T-Cadinol	nd–200	—
Humulenol	1–1150	500, 2500
Humuladienone	nd–43	—
Humulol	nd–220	—
Clovanediol	51–677	—

*nd: not detectable. #threshold determined in water.

bitter isomerised counterparts, it does permit the extraction, steam frac-
tionation and chemical modification of part of the hop oil present. Late
hop flavour in lager beers is often described as floral and spicy, consisting
primarily of monoterpene alcohols such linalool and geraniol, together
with the nor-carotenoids and cyclic ethers. For the production of ales, it is
common practice to add whole hop cones or whole hop pellets to the
cask. The resulting dry hop character is relatively simple, comprising of a
combination of monoterpenes such as myrcene, aliphatic esters and
linalool.

Whilst traditionally beer was produced with whole hop cones, this
practice is relatively rare today, with the use of standardised pellets and
extracts which can provide bitterness, hoppy flavour, or both and highly
refined extracts for the modification of hoppy flavours at the end of beer
production. Finally, hoppy aroma is a hop variety characteristic. Thus
while hops for providing bitterness are sold on the basis of their α-acid
content, specific varieties prized for the quality of their aroma (Table 3.12)
often command premium prices. The current status of research however
means that the ability of a hop variety to impart a good aroma cannot
readily be measured analytically and still requires brewing trials and
sensory evaluation to ensure that this is indeed the case.

Malt Flavours

Malt is the major raw material for most beers produced world-wide. The
vast majority of this is white malt, which is so-called because it is kilned

Table 3.12 *Classification of some common hop varieties used for global beer production. This classification is somewhat arbitrary as all varieties contain both bitter and oil components*

Classification	Variety	Country of origin
Bitter (high α-acid content)	Brewers Gold	UK
	Galena	USA
	Magnum	Germany
	Nugget	USA
	Pride of Ringwood	Australia
	Wye Target	UK
Aroma hops (confer good quality hop aroma on beer)	Fuggles	UK
	Goldings	UK
	Hallertau Mittelfrüh	Germany
	Hersbrucker	Germany
	Lublin	Poland
	Mount Hood	USA
	Saaz	Germany
	Willamette	USA

primarily to remove the water absorbed during germination, rather than to generate colour. The act of kilning also helps to reduce the levels of flavour-negative green notes which are characteristic of unkilned malt, due to the presence of aldehydes such as hexanal. The degree of colour formation during kilning correlates to the formation of increasing amounts of flavour-active Maillard reaction products, which are inevitable given that the main aim of the malting process is to release fermentable carbohydrates and free amino nitrogen for subsequent fermentation. Thus the use of speciality malts, which are applied in smaller proportions, add both colour and flavour to the final beer (Table 3.13).

The heat-promoted chemical reactions which occur during malt kilning are complex, including the thermal degradation of phenolic acids, the caramelisation of sugars, Maillard reactions (including the Strecker degradation of amino acids), and thermal degradation of oxygenated fatty acids derived both chemically and enzymically from lipids. Thus a range of volatile compounds are formed, such as fatty acids, aldehydes, alcohols, furans, ketones, phenols, pyrazines and sulfur compounds. The presence of these compounds in malt indicates that they are likely to occur in the sweet wort. However, many are lost during the boiling stage either by evaporation, chemical breakdown or, during fermentation by the action of yeast. Of those that survive into beer, DMS is one of the most significant particularly for lager beers (see Sulfur Compounds).

A significant component of malt is lipid – typically 4% (w/w) of the dry grain. Most of this material, being relatively insoluble in aqueous media, is lost with the spent grains or precipitated with the solids or trub which

Table 3.13 *Typical colour specifications and flavour attributes for malts and roasted barley. The darker (i.e. higher EBC value) malts tend to be used in darker or speciality beers*

Malt	Colour (°EBC)	Typical flavour attributes
Pale ale	4.5–4.8	Biscuity
Caramalt	25–35	Sweet, nutty, cereal, toffee
Crystal	100–300	Malty, toffee, caramel
Amber	40–60	Nutty, caramel, fruity
Chocolate	900–1200	Mocha, treacle, chocolate
Black	1250–1500	Smoky, coffee
Roasted barley	1000–1550	Burnt, smoky

are formed both during the boil and the subsequent cooling operations. The major lipid component is linoleic acid, which can yield many C_5–C_{12} saturated and unsaturated aldehydes, ketones and alcohols, some of which are highly flavour-active (see Chapter 4).

A discussion of the detailed chemistry of Maillard reactions is beyond the scope of this text. For the purposes of this chapter, it is worth noting that Maillard reactions result in the formation of a large number of oxygen heterocyclic compounds, such as furans, furanones and pyrones. Compounds such as furaneol, maltol and isomaltol (Table 3.14) are highly flavour-active and can contribute to beer flavour. During the course of Maillard reactions, the oxygen of furan rings may be replaced by sulfur or nitrogen moieties, leading to the formation of the corresponding thiophenes and pyrroles. Other malt-derived heterocycles identified in malt include thiazoles, thiazolines, pyridines, pyrrolizines and pyrazines. Some pyrazines, such as the dimethylpyrazines, can occur at levels above two flavour units in some cases, thereby substantially affecting beer flavour. These compounds have been variously described as malty, oxidised and sweet.

Other Contributors to Beer Flavour

There are other compounds which have not yet been considered. One important compound is 4-vinylguaiacol, which is produced by the yeast-induced decarboxylation of ferulic acid (Figure 3.15). Unlike ferulic acid, 4-vinylguaiacol has a potent smoky, clove-like aroma and indeed is readily detected in certain products, such as wheat beers. Given that the flavour threshold of 4-vinylguaiacol is $300 \, \mu g \, l^{-1}$ and that it can occur in beer typically at levels of 50 to $550 \, \mu g \, l^{-1}$, this compound can significantly alter beer flavour. Its presence can also indicate that the beer is contaminated with wild yeast strains.

Table 3.14 *Significant flavour-active oxygen heterocycles found in malt and beer*

Compound	*Structure*	*Flavour attributes*
Furaneol	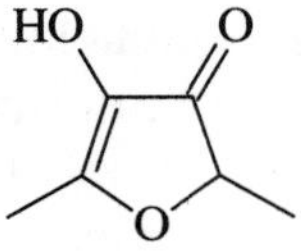	Toffee, caramel
Maltol	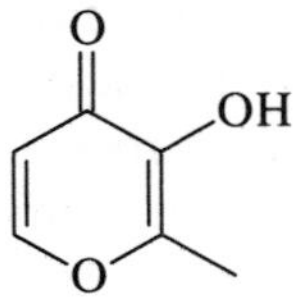	Caramel
Isomaltol	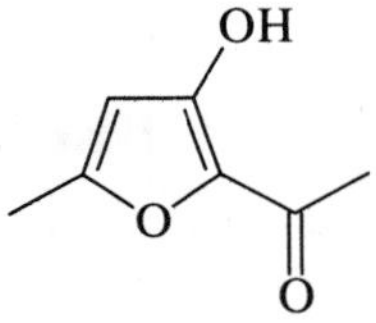	Caramel, burnt sugar

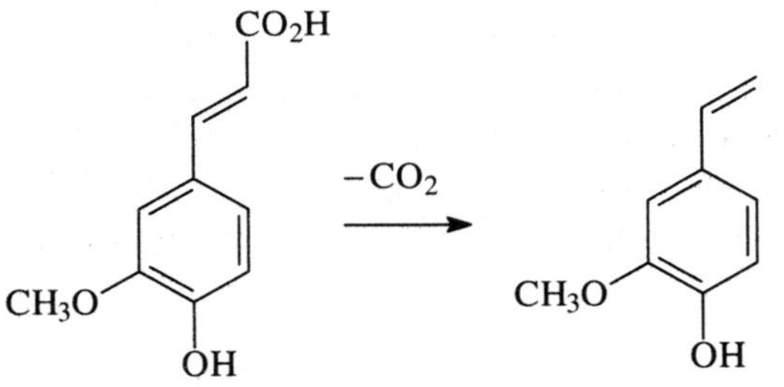

ferulic acid 4-vinylguaiacol

Figure 3.15 *Formation of 4-vinylguaiacol from* trans-*ferulic acid*

DRINKABILITY

The phenomenon known as drinkability is well-recognised within the brewing industry but, as yet, has not been fully defined. It appears to be an interplay between sensory, physiological and psychological factors. Recent studies have shown that there is a high degree of negative correlation between the desire to drink and perceived stomach fullness, the latter confirmed by ultrasonic measurement of the cross-sectional area of the pylorus antrum.[8] In turn drinkability correlated with positive appraisal of beer taste. In a separate study, Guinard *et al.*[9] found that the only positive factors corresponding to beer thirst-quenching, drinkability and refreshing characteristics were degree of carbonation and bubble density. In contrast, this investigation found significant negative contributors to the drinkability sensation, several of which appeared to be relate to components which prolong the aftertaste of beer (*i.e.* bitterness, astringency, burnt, viscous). The understanding of the role of ethanol as an energy source for human metabolism still appears contradictory and could also play an important role in the human regulation (or otherwise) of beer consumption.

THE MOUTHFEEL OF BEER

The mouthfeel of beer is undoubtedly important to the consumer, but this has proved a difficult parameter to describe analytically. Considered to be related to texture, the sensation is relatively short-lived and of lower intensity than comparable sensations from foodstuffs. There is some debate about how the term mouthfeel ought to be described. This comes about partly because mouthfeel can often be confused with flavour characteristics. A number of sets of terms have been proposed in the literature in an attempt to describe mouthfeel sensorially. Viscosity-related terms are often used, indicating that these most easily recognised by a panel of assessors. The classification proposed by Langstaff *et al.*[10] uses nine attributes arranged into the three groups of carbonation, fullness and mouthfeel (Table 3.15).

SENSORY ASSESSMENT OF BEER

The extent to which the appraisal of the sensory characteristics of beer has occupied the minds of brewing scientists is evident from the large body of literature which has been compiled on the subject. Perhaps the best-known product of detailed studies on beer flavour is the construction of the beer flavour wheel (Figure 3.16),[10] which, as mentioned above, has since had one modification suggested to incorporate aspects of texture and mouthfeel.[11] Whilst many recognise the limitations of the flavour wheel, it does nonetheless bring together the main descriptors associated with many beers produced commercially. In parallel with related industries, sensory evaluation techniques can be classified accord-

Table 3.15 *Mouthfeel descriptors as defined by Langstaff et al.*[10]

Group	Attributes
Carbonation	Sting
	Bubble size
	Foam volume
	Total carbon dioxide
Fullness	Density
	Viscosity
Afterfeel	Oily mouthcoat
	Astringency
	Stickiness

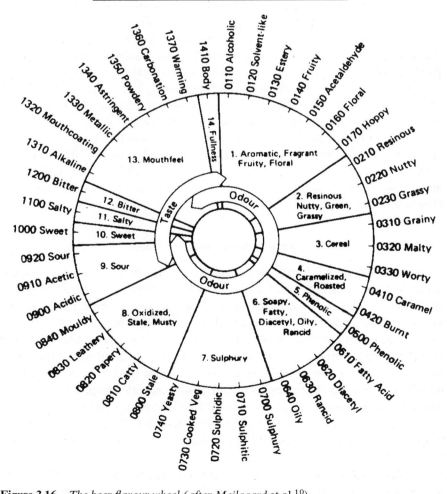

Figure 3.16 *The beer flavour wheel (after Meilgaard et al.*[10])

ing to the information sought and the way in which the sensory experiment is set up.

'Difference tasting' procedures are those which are designed to compare and contrast the flavours of two or more products. The aim of such a test is to establish whether there are differences in flavours between the test products which are discernible by a substantial proportion of the population. Thus, the number of sensory assessors needs to be such that there is sufficient data for robust statistical analysis. Careful attention to the conditions under which the test is carried out is important to minimise the effect of other non-flavour variables (*e.g.* visual cues, background distractions) on the assessors.

'Descriptive testing' generally requires more training of the assessors, as they are requested to try and quantify the levels of each of a number of flavour attributes. Most brewing companies have tailored such tests for their own products, but it is important to continually monitor assessors' performance to ensure that such tests are as robust as possible. Such checking of sensory panels is now facilitated by the availability of software packages which can not only gather data electronically, but also use this data to automatically track assessors' performance.

Other more specialist tests include 'time-intensity assessment', first described in the 1950s, which is particularly useful for long-lived flavour attributes such as bitterness and astringency. Such tests are again challenging for the assessor, but can yield useful information on the temporal characteristics of beers.

Understanding Sensory Data

As data sets become more complex and with the need to relate disparate groups of information – such as analytical, sensory and consumer data – multivariate statistical techniques have become an essential tool for the sophisticated sensory analyst (see Meilgaard[12] for review). While a detailed discussion of such methods is beyond the scope of this text, it is worth pointing out that data from descriptive tests is often highly correlated and thus is not readily amenable to more traditional methods of analysis such as multiple regression. Thus principal component analysis, principal component regression and partial least squares regression, which are designed to cope with highly correlated data, have become popular as desktop computer power and software become more readily available.

As an example of such an application, consider the descriptor 'alcoholic' for beer. Of course, given the levels of ethanol present in beer, this would be expected to be the major contributor to this descriptor. However, there is little correlation between alcohol content as determined by analysis and the sensory perception of alcoholic flavour in beer. In our laboratory, partial least squares regression analysis of esters and alcohols in beer (the *X*-matrix) *vs* alcoholic flavour (*Y*-vector) clearly demonstrated the significant contributions of other volatiles, particularly higher alcohols, to alcoholic flavour. The point is that there is rarely a one-to-one mapping of one flavour attribute with one analytically determined compound and therefore such numerical tools are often called upon to extract information from sensory data.

BEER FLAVOUR

The components described above all combine to give rise to what consumers know to be beer flavour. The sheer number of effects and components which contribute to this is clearly immense, which goes a long way to explain why description of beer flavour, and consumer responses to beer flavour are, on the whole, extremely challenging to predict *a priori*. It is important to appreciate that study of the compounds responsible for beer flavour have been firmly in the domains of organic and analytical chemistry. These disciplines require substantial simplification of the problem – achieved by dissecting beer to the level of single compounds or groups of compounds. This is the antithesis of sensory perception which is a holistic experience of product attributes simultaneously. Meilgaard has elegantly summarised the major flavour attributes of beer according to the typical number of flavour units at which they exist in beer (Table 3.16). Clearly, carbon dioxide, the hop bitter acids and ethanol all play the major role in definition of beer flavour, but other compounds, resulting either from defects or because they are specific characters associated with certain products, can be present at levels far in excess of two flavour units.

While a substantial body of literature on the sensory aspects of beer has developed, the same cannot be said of consumer information. This is no doubt due in part to the proprietary nature of such information, as it often pertains to specific brands. Also, consumers do not react solely to flavour stimuli. Tasting a food or drink is one of the final stages of a consumer's interaction, and there are other factors, such as visual impact of the product (see Chapter 2) and, even before that, the impact of marketing, cost and social pressures all affect consumer responses. Understanding the behaviour of consumers and the part that flavour has to play on consumer behaviour would be a highly fruitful area of research, and one which is likely to become more prevalent in the future.

Table 3.16 *Typical compounds responsible for beer flavour. Compounds present at < 0.1 flavour units are considered to account for less than 30% of total beer flavour*

Number of flavour units	Principal constituents	
	most beers	niche and defective beers
>2 (*changes in levels of these compounds substantially alter beer flavour*)	Carbon dioxide Ethanol Iso-α-acids	Modified iso-α-acids Humuladienone *trans*-2-Nonenal Diacetyl/pentanedione Hydrogen sulphide Dimethyl sulphide 3-Methyl-2-butene-1-thiol Acetic acid Acetaldehyde Iron, copper ions
0.5–2 (*changes in levels of these compounds cause a small flavour change*)	Isoamyl acetate Ethyl hexanoate Isoamyl alcohol Fatty acids (C_6–C_{10}) Ethyl acetate Butyric acid 3-Methylbutanoic acid Phenylacetic acid Polyphenols Non-volatile acids Other hop compounds	Modified iso-α-acids Dimethyl sulphide Humuladienone Acetaldehyde Acetic acid
0.1–0.5 (*changes in level of any of these does not alter beer flavour*)	2-Phenylethyl acetate 3-Methylbutanal 4-Thiapentanal Acetoin 2-Aminoacetophenone γ-Valerolactone 4-Ethylguaiacol	

SUMMARY

The deceptively simple term 'flavour' incorporates the experiences of the consumer when eating or drinking. An indication of the complexity of beer flavour is that it can be crudely and incompletely resolved into taste, aroma and mouthfeel. A large body of literature has gone some way to understanding the cause-and-effect relationships between beer composition and sensory experience, but there is still too much to do, not only in a chemical-analytical sense but also the aspects of sensory signal transduction and the neural processing that results in a consumer's holistic sensory perception. For many beers, bitterness, alcohol and carbon dioxide dominate the sensory profile of beer, and are characters readily perceived by the consumer. Other characters, such as hoppy, fruity/estery, dimethyl sulfide and sufury, contribute to brand differentiation. Finally, compounds such as diacetyl and aliphatic aldehydes (see Chapter 4) are often indicative of defective beers.

REFERENCES

1 M.C. Meilgaard, 'Flavor chemistry of beer: Part II: Flavor and threshold of 239 aroma volatiles', *Master Brew. Assoc. Am. Tech. Q.*, 1975, **12**, 151–168.

2 F.L. Rigby, 'A theory on the hop flavor of beer', *Proc. Am. Soc. Brew. Chem.*, 1972, 46–50.

3 P.S. Hughes and W.J. Simpson, 'Bitterness of congeners and stereoisomers of hop-derived bitter acids found in beer', *J. Am. Soc. Brew. Chem.*, 1996, **54**, 234–237.

4 M. Moir, 'Hop aromatic compounds', *Proceedings of the EBC Symposium on Hops*, 1994, 165–180.

5 P.K. Hegarty, R. Parsons, C.W. Bamforth and S.W. Molzahn, 'Phenylethanol – a factor determining lager character, *Proc. Eur. Brew. Conv. Cong.*, 1995, 515–522.

6 P. Semmelroch and W. Grosch, 'Study on the coffee impact odourants of coffee brews', *J. Agric. Food Chem.*, 1996, **44**, 537–543.

7 M. Laurent, B. Geldorf, L. van Nedervelde, S. Dupire and A. Debourg, 'Charactertization of the aldoketoreductase system involved in the removal of wort carbonyls during fermentation', *Proc. Eur. Brew. Conv. Cong.*, 1995, 337–344.

8 Y. Nagao, H. Kodama, T. Yonezawa, S. Fujino, K. Nakahara, K. Haruma and T. Fushiki, 'Correlation between the drinkability of beer and gastric emptying', *Biosci. Biotech. Biochem.*, 1998, **62**, 846–851.

9 J-X. Guinard, A. Souchard, M. Picot, M. Rogeaux and J-M. Sieffermann, *Appetite*, 1998, **31**, 101–115.

10 M.C. Meilgaard, C.E. Dalgliesh and J.F. Clapperton, 'Beer flavor terminology', *J. Am. Soc. Brew. Chem.*, 1979, **37**, 47–52.

11 S.A. Langstaff, J-X. Guinard and M.J. Lewis, 'Sensory evaluation of the mouthfeel of beer', *J. Am. Soc. Brew. Chem.*, 1991, **49**, 54–58.

12 M.C. Meilgaard, 'Current progress in sensory analysis. A review', *J. Am. Soc. Brew. Chem.*, 1991, **49**, 101–109.

Chapter 4

Maintenance of Beer Quality

INTRODUCTION

From what has gone before, it is clear that beer is a complex mixture of a diverse range of components in an acidic medium. Thus it is perhaps not surprising that the maintenance of beer quality throughout its lifetime is a considerable challenge. Market forces add additional weight to this challenge with increasing demands for extended (often international) distribution chains and the drive towards ever longer shelf lives. Here, four aspects of beer stability will be discussed: the change in beer flavour, the loss of foam stability, the formation of particles and microbial contamination.

BEER FLAVOUR STABILITY

The flavour changes which occur in beer over time are well-recognised and have a long and vigorous history of research. Brewers strive to give their consumers beer of high and consistent quality, so flavour changes that occur over time threaten to compromise such consistency. The problem does not yet appear to have a holistic solution, which is perhaps indicative of the wide range of possible chemical changes which can occur in beer over time. From a chemical point of view, it is pertinent to remember two facts: firstly that a period of months is a long reaction time compared to lab-scale experiments so that chemistry that might appear unfeasible on paper may in fact occur to a significant extent in beer, albeit at very slow rates. Secondly, whilst it is inevitable that chemical changes will occur in beer with time, it is the formation of flavour-active components which is of immediate concern from a flavour stability perspective. So for example, the formation of *trans*-2-nonenal *via* trace lipid oxidation is primarily a flavour stability problem because it has a flavour threshold of around $100 \, \text{ng} \, l^{-1}$ (700 pM).

There are several possible routes to flavour-active breakdown products in beer. Each of these will be discussed below. It should be remembered though that some of these issues still invoke fierce debate and that one of the few areas that most agree on is the detrimental role of oxygen on beer flavour stability.

Potential Sources of Flavour Instability

The Role of Oxygen

Oxidative degradation of beer is generally considered to be the major source of beer flavour instability. Thus the control of oxygen in the brewing process has long been a preoccupation for brewers. Oxygen itself is not particularly reactive, existing as a triplet ground state. However, it can be activated by a number of agents, such as transition metal ions (*e.g.* copper, iron), enzymes, photosensitisers and light, to give singlet oxygen.[1] This is a much more reactive compound, and is the precursor to a family of so-called reactive oxygen species (ROS, Figure 4.1). Which of these ROS are the most damaging is a matter for debate. Thus, although the hydroxyl radical is generally much more reactive than the hydroperoxyl radical, the former will react with less specificity than the latter, so that it is potentially a less effective flavour-destabilising agent.

Beer rapidly reacts with any molecular oxygen to which it is exposed, so for most mainstream beers, after the end of fermentation, it is important to minimise any exposure to oxygen during downstream processing – in particular during filtration and packaging. In this context, it is instructive to consider antioxidant activities. Beer generally contains a range of antioxidants, the most important of which are the polyphenols, phenolic acids (Figure 4.2) and sulfur dioxide (see later). The profile of polyphenols changes substantially during beer production. This is due to their ability to precipitate beer polypeptides, which aids in conferring physical stability on beer (see later), and their sensitivity to oxidation. So whilst beer contains ample antioxidant components to at least partially protect it against oxidative damage, this pool is finite and needs to be protected in the pack to maximise their effectiveness.

An alternative approach for the prevention of oxidative damage is the use of oxygen-scavenging crown seals in bottle crown caps. It is not immediately apparent that oxygen can diffuse across crown seals from the atmosphere, but whilst this is against a pressure gradient, it is merely a response to the oxygen partial pressure gradient. Such scavengers, which

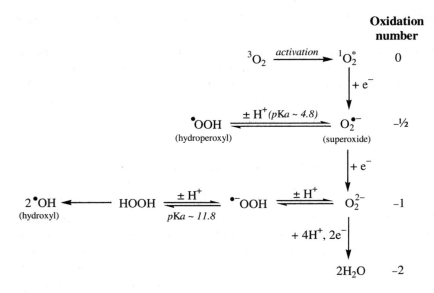

Figure 4.1 *Schematic showing the potentially damaging reactive oxygen species (ROS) which can be generated from molecular oxygen. Of these, the hydroxyl radical (˙OH) and hydroperoxyl (˙OOH) are considered to be the most reactive*

are incorporated into the crown liner material, tend to be based on sulphite and/or ascorbate, isoascorbate or erythrobates. The use of transition metals catalyses the rate of the reactivity with oxygen.

It is not possible to eliminate oxygen totally from the brewing operation. This is because oxygen is required for the malt production and also for aerobic yeast growth at the beginning of fermentation. During mashing-in, there are opportunities to reduce levels of oxygen pick-up, by deaeration of the water and milled malt, filling the mash tun from the bottom to minimise turbulence and keeping the hatches to the mash tun closed during the mashing-in stage. There are also those who advocate a higher mashing-in temperature to deactivate enzymes such as lipoxygenase more rapidly. However, it should be remembered that although enzymes are more rapidly denatured at higher temperatures, they also turn-over faster, so that the possibility of oxidative damage even at higher mashing-in temperatures cannot be ruled out.

Lipid Oxidation

In common with other foodstuffs, lipid oxidation is considered to be a significant source of beer flavour instability. There is generally little lipid in beer, but during beer production, specifically in the malting and

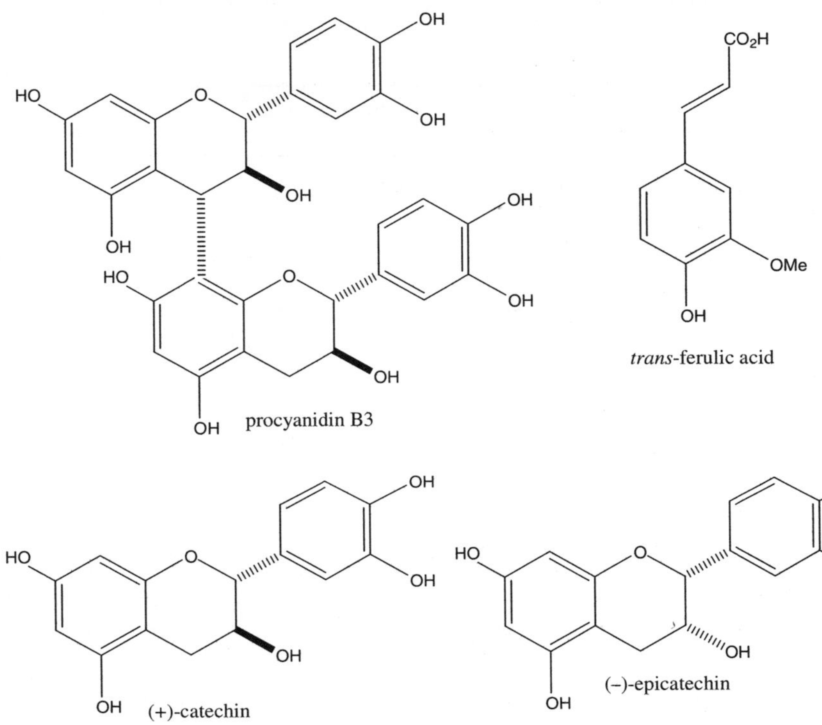

Figure 4.2 *Representative antioxidant compounds found in beer. The polyphenols found in beer and its raw materials are by no means completely defined, and consist not only of 'monomeric' polyphenols such as catechin but also dimers, trimers and higher oligomers*

mashing operations, the lipid levels in malt – typically $45 \, \text{g kg}^{-1}$ – and the presence of malt lipoxygenases, provide scope for the enzymic oxidation of malt lipid components. The major compound present in the malt lipid fraction is linoleic acid (9,12-octadecadienoic acid; Table 4.1) which contains a reactive stepped diene moiety. This is readily attacked by lipoxygenases (iron-containing dioxygenases) to give either the 9- or the 12-hydroperoxylinoleic acid (Figure 4.3). These products are reactive in their own right, yielding a range of oxygen-containing compounds (Figure 4.4). Clearly lipid oxidation is a complicated process and for beer is considered to be both enzymic and non-enzymic.

Linoleic acid is by no means the only lipid in malt. Other free fatty acids, such as linolenic acid (9,12,15-octadecatrienoic acid) also occur in malt lipids. As this compound contains two active methylene carbons it is perhaps unsurprising that it too is sensitive to oxidative degradation.

Table 4.1 *Predominant fatty acids occurring in malt lipids. Note the preponderance of unsaturated lipids*

Fatty acid	Proportion of total (%)
Palmitic acid (16:0)	26
Stearic acid (18:0)	1
Oleic acid (18:1)	10
Linoleic acid (18:2)	57
Linolenic acid (18:3)	6

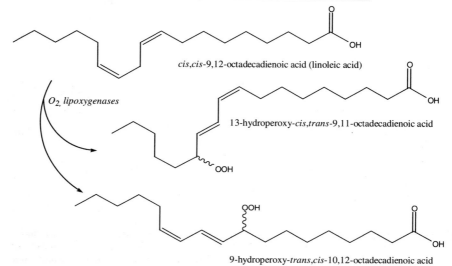

cis,cis-9,12-octadecadienoic acid (linoleic acid)

O_2, *lipoxygenases*

13-hydroperoxy-*cis,trans*-9,11-octadecadienoic acid

9-hydroperoxy-*trans,cis*-10,12-octadecadienoic acid

Figure 4.3 *Lipoxygenase-induced oxidation of linoleic acid. The two hydroperoxides are generated from two isoforms of lipoxygenase found in malt. Of these, the 9-hydroperoxy compound is considered to be the most damaging as it is an intermediate for the formation of the highly flavour-active aldehyde* trans-2-nonenal. *(Note: for the brewing case, configuration of the double bonds is a matter for conjecture, as is the chirality of the peroxy-bearing carbon.)*

Hop Acid Degradation

The hop bitter acids are ubiquitous in all beers sold today. Their properties with regard to foam stabilisation (Chapter 2) and bitterness (Chapter 3) have been discussed previously. However, they are chemically labile, and steadily degrade during beer storage. Thus when irradiated with light, hop acids rapidly release the highly aroma-active mercaptan, 3-methyl-2-butene-1-thiol (MBT), as discussed previously (Chapter 3). However, other compounds, such as methyl ketones, aldehydes and

5-methyl-4-pentenoic acid can also be generated, at least in model systems.[2,3] In this regard, it is interesting to note that at this time, tetrahydroiso-α-acids were reported to give rise to a beer of enhanced flavour stability relative to a conventionally bittered control.

Strecker Degradation of Amino Acids

Finished beer contains small quantities of a range of amino acids and sugars which can undergo a series of non-enzymic browning reactions during malt kilning and wort boiling. Crucially, such a series of reactions can ultimately yield aldehydic off-flavours (Figure 4.5). These reactions can be classified into three stages:

- Initialisation – condensation of sugar with amino compounds (mainly amino acids) followed by Amadori or Heyns rearrangements.
- Formation of intermediates – formation of flavour compounds such as O-heterocycles, mono- and dicarbonyl compounds, Strecker aldehydes, pyrroles and N-heterocycles.
- Polymerisation – formation of melanoidins from intermediates.

The presence of a Strecker aldehyde in malt and wort is accompanied by its corresponding alcohol, presumably as a result of enzymic reductases. The presence of such higher alcohols though, has been suggested by some to be a potential source of staling aldehydes by way of their oxidation during beer storage. There is also increasing evidence for the presence of active keto- and aldoreductase activity during fermentation, which while again there is a pool of higher alcohols being formed, they could be an ultimate source of stale flavour. It is fair to say though that there is some debate as to the contribution of Strecker degradation to beer flavour instability. Some authors consider that Strecker degradation can only occur at elevated temperatures (*i.e.* above 80 °C) whilst others have shown that this reaction can occur at temperatures as low as 50 °C.

Melanoidin-mediated Oxidation of Alcohols

Melanoidins are poorly defined colour bodies that occur in all beers to a greater or lesser extent. They are the ultimate products from the Maillard reaction (Figure 4.5(b)). The paucity of knowledge stems from the complex pathway of the Maillard process and the sheer number of reactants (amino acids, sugars) which are present both during malt kilning and wort boiling. Experiments with model systems have indicated that melanoidins can take part in a redox reaction with higher alcohols, acting

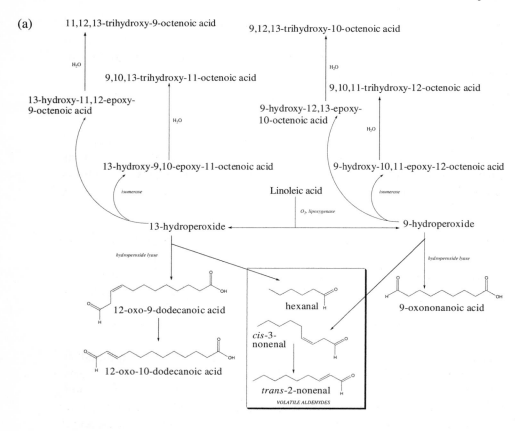

Figure 4.4   *(a) Subsequent reaction products of linoleic acid hydroperoxides. As well as the formation of volatile flavour-active aldehydes, a range of oxygenated fatty acids can be formed. Some of these, particularly the trihydroxy fatty acids can persist in the beer by virtue of their higher aqueous solubility and may be a source for the generation of aldehyde off-flavours in pack. (b) Structures of linoleic acid oxidation products*

as a hydrogen atom acceptor (Figure 4.6). Interestingly, one of the few papers in this area considers that only low molecular weight melanoidins are involved in this reaction.

   There are other, even less well-defined reactions which melanoidins are thought to take part in, including other redox reactions with polyphenols and carbonyl compounds.

Carbonyl Compounds in Beer

From what has been said above, it is clear that whilst carbonyl compounds, especially aldehydes, are not the sole cause of beer flavour

(b)

11,12,13-trihydroxy-9-octenoic acid

9,10,13-trihydroxy-11-octenoic acid

13-hydroxy-11,12-epoxy-9-octenoic acid

13-hydroxy-9,10-epoxy-11-octenoic acid

instability, they play a primary role. Not all aldehydes and ketones in beer are formed as a result of beer flavour deterioration, although some are relatively reactive and therefore their levels are prone to change. Presumably this can be due to oxidation, reduction, or Aldol condensation. A fourth route of loss, at least for the α,β-unsaturated aldehydes, is the addition of bisulfite to the carbon–carbon double bond (see below). The carbonyl compounds found in beer are summarised in Table 4.2.

Sulfur Dioxide

Above all other compounds, the behaviour of sulfur dioxide has exercised the minds of the international brewing science community. It is ubiquitous in beer, being produced by yeast during fermentation and, on occasions, added during the brewing process. Sulfur dioxide has three forms in aqueous solution ($SO_2.H_2O$, HSO_3^-, SO_2^{2-}). At beer pH though, the predominant species is the bisulphite anion, with significant levels of free sulfur dioxide only being formed when the pH is below 3. It is well-known to form adducts with aldehydes (Equation 4.1), the α-hydroxysulfonates. These adducts, in contrast to the free aldehydes, are

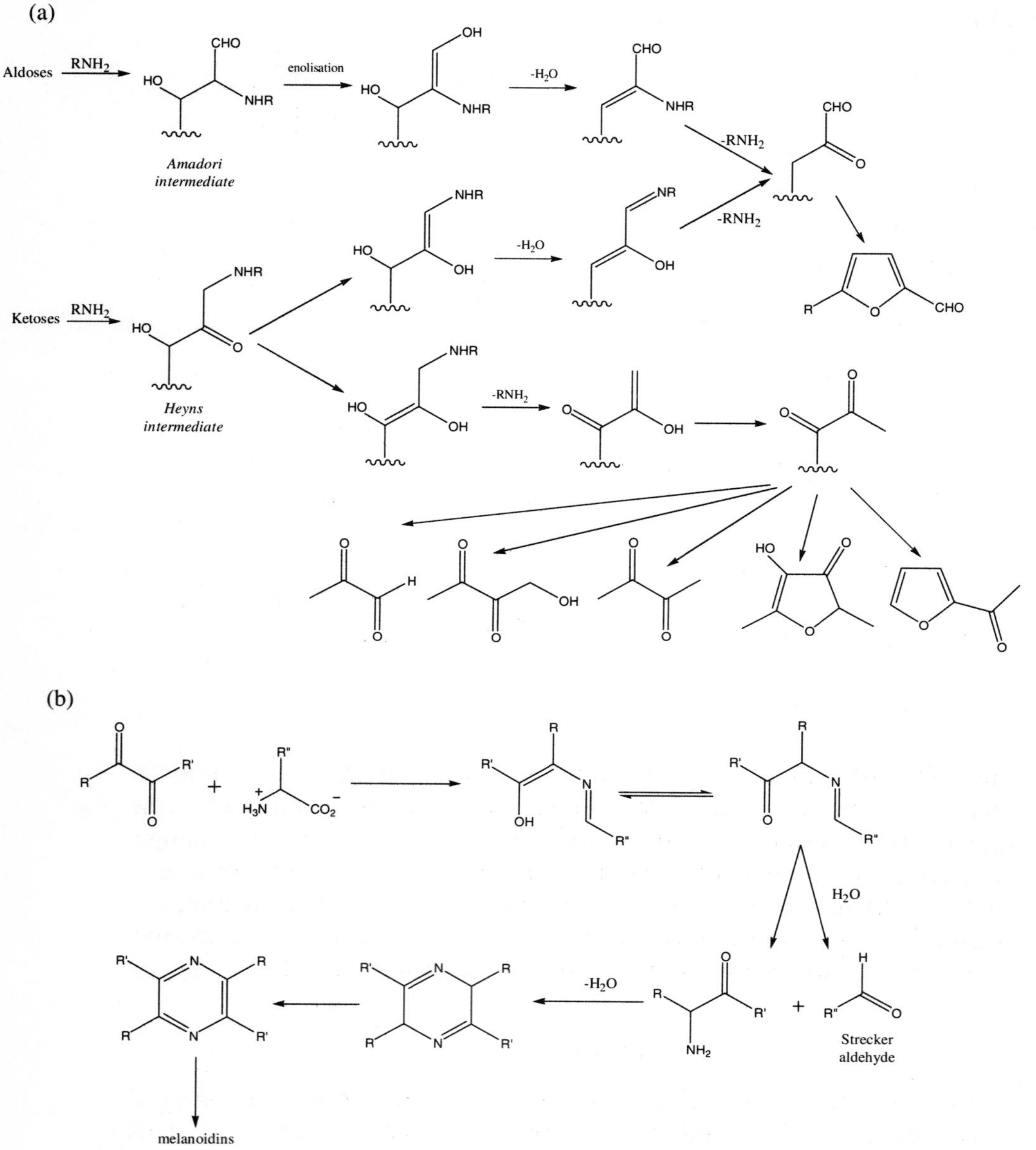

(a)

(b)

Figure 4.5 *Representative reaction scheme showing the steps for (a) the formation of aldehydes from sugars and amino acids and (b) subsequent formation of pyrazines and melanoidins*

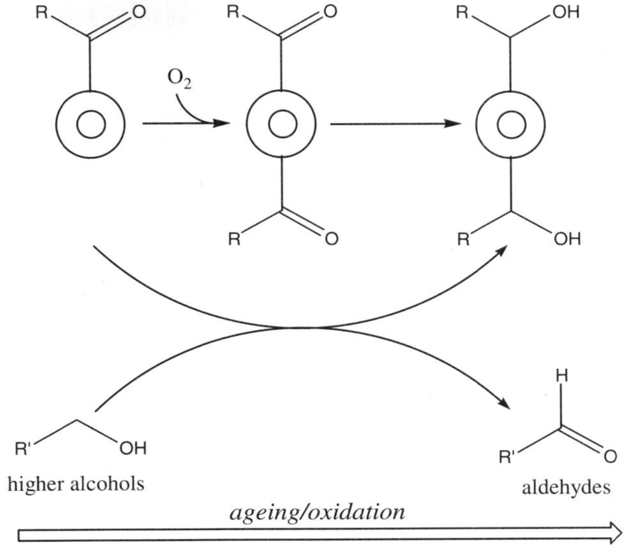

Figure 4.6 *Melanoidin-mediated oxidation of higher alcohols. Melanoidins are thought to act as hydrogen atom acceptors, oxidising higher alcohols in the process. Whilst the presence of oxygen is not mandatory, it does appear to accelerate the formation of more reactive melanoidin species*

relatively involatile and have a higher aqueous solubility. So, in fact, the formation of such adducts should mask any aldehydes present and indeed this has been shown to be the case. The efficacy of binding of bisulfite to carbonyl compounds does vary. Increased steric hindrance around the carbonyl interferes with bisulfite binding, so that aldehydes bind to bisulfite to a greater extent (usually 70–90% in a 1:1 mixture) than ketones (methyl ketones usually give yields of 12–56% in a 1:1 mixture). Of course, the reaction is reversible, so that the presence of an excess of bisulfite will increase yields of the adduct.

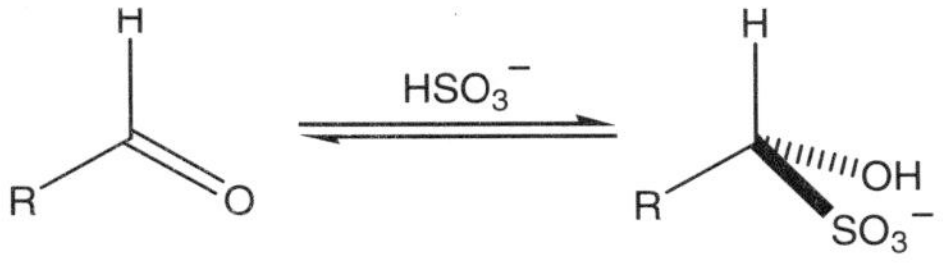

(4.1)

Nevertheless, sulfur dioxide is lost during beer storage with pseudo-first order kinetics, so that it has been observed to have a half-life of around 100 days, although extremes of one week and three years have also been noted. Whilst the fate of this sulfur dioxide in beer is unknown, recent reports indicate that unsaturated aldehydes can form a variety of irreversible adducts with sulfur dioxide[4,5] (Figure 4.7). In other foodstuffs,

Table 4.2 Summary of aldehydes and ketones found in beer. Many occur substantially below their flavour threshold, although additive and synergistic sensory effects cannot be ruled out. All data is based on Meilgaard[4] except where indicated

Compound	Typical levels in beer (mg l⁻¹)	Threshold (mg l⁻¹)	Flavour units	Possible source(s)*	Flavour
Aldehydes					
Acetaldehyde	2–20	25	0.08–0.8	1	Green, paint
Propanal	0.01–0.3	5, 30	0.00–0.06	1?	Green, fruity
Butanal	0.03–0.2	1.0	0.03–0.20	1?	Melon, varnish
trans-2-Butenal	0.003–0.02	8.0	0.00	2??	Apple, almond
2-Methylpropanal	0.02–0.5	1.0	0.02–0.5	3	Banana, melon
C₅ Aldehydes	0.01–0.3	ca. 1.0	0.01–0.3	3	Grass, apple, cheese
Hexanal	0.003–0.07	0.35	0.01–0.02	4	Bitter, vinous
trans-2-Hexenal	0.005–0.01	0.6	0.01–0.02	2, 4	Bitter, astringent
Heptanal	0.002	0.075	0.03	4	Aldehyde, bitter
Octanal	0.001–0.02	0.04	0.03–0.5	4	Orange peel, bitter
Nonanal	0.001–0.011	0.018	0.06–0.6	4	Astringent, bitter
trans-2-Nonenal	0.00001–0.002	0.0001	0.09–18	2, 4	Cardboard
cis-3-Nonenal**	—	0.0005	—	4	Soy-bean oil
trans-2-cis-6-Nonadienal	—	0.00005	0.0–0.5	4	Cucumber, green
Decanal	0.0–0.003	0.006	0.0–0.5	4	Bitter, orange peel
Decadienal	—	0.0003	—	4	Oily, deep fried
Furfural	0.01–1.0	200	0.0	5	Papery, husky
5-Methylfurfural	<0.01	17	0.0	5	Spicy
5-Hydroxymethylfurfural	0.1–20	1000	0.0–0.02	5	Aldehyde, stale
Ketones					
3-Methylbutan-2-one	<0.05	10	0.0	3	Ketone, sweet
3-Methylpentan-2-one	0.06	—	—	3	—
4-Methylpentan-2-one	<0.013	1.5	<0.01	3	—
3,3-Dimethylbutan-2-one	—	—	—	3	—
6-Methyl-5-hepten-2-one	0.05	—	—	3	—

Heptan-2-one	0.04–0.11	2.0	0.0–0.06	3	Varnish, hops
Octan-2-one	0.01	0.25	0.04	3	Varnish, walnut
Nonan-2-one	0.03	0.20	0.15	3	Ketone, varnish
Decan-2-one	—	—	—	3	Ketone, flowery
Undecan-2-one	—	—	—	3	Ketone, green plant
Oct-1-en-3-one***	—	0.0001	—	4	Metallic, mushroom
Octa-1-*cis*,5-dien-3-one***	—	1×10^{-6}	—	4	Metallic, geraniums

*Sources of carbonyl compounds:
1. Lower alcohol oxidation
2. Aldol condensation
3. Hop-derived
4. Fatty acid oxidation
5. Maillard product
**Data from Swarboda & Peers[5]
***Data from Sakuma & Kowaka[6]

sulfur dioxide has been observed to react with a variety of components, some of which occur in beer[6] (Figure 4.8). One can speculate that this is due to the occurrence of chelotropic reactions, sulfonation of free thiols and redox reactions with polyphenols and quinones.

Free Radicals and Beer Flavour Instability

From what has been said above, it is apparent that free radicals could play a role in the flavour instability of beer. Indeed, a survey of the recent brewing science literature reveals the increasing use of electron spin resonance (ESR) in well-equipped brewing laboratories.[7] The best-known published application is the measurement of the 'lag time'. Here, a spin-trap, typically PBN (*N*-*t*-butyl-α-phenylnitrone), is added to a beer sample and the whole sample is heated, typically at 60 °C. Initially there is no change in the total free radical concentration in the sample as measured by ESR. Eventually the free radical signal begins to rise linearly with time (Figure 4.9). The time taken for the signal to become time-dependent is termed the lag time and has been shown to correlate with a range of parameters, such as soluble iron. Bench-top ESR instruments are now installed in some breweries to maintain a check on potentially damaging shortening of lag times. Which radicals the spin-trap bind to is unknown. One recent publication points to the primary hydroxyethyl radical, de-rived by abstraction of a hydrogen atom from the methyl group of ethanol (Equation 4.2).

$$(4.2)$$

Distortion of Beer Flavour

Whilst there is little doubt as to the deleterious role which compounds such as staling aldehydes have on beer flavour instability, there are other components too which should be considered. Dalgliesh[8] cited a number of sensory attributes which change during beer storage:

- a decrease in bitterness
- an increase in sweetness
- an increase in diacetyl
- the formation of a ribes or catty off-flavour which is eventually lost during extended storage.

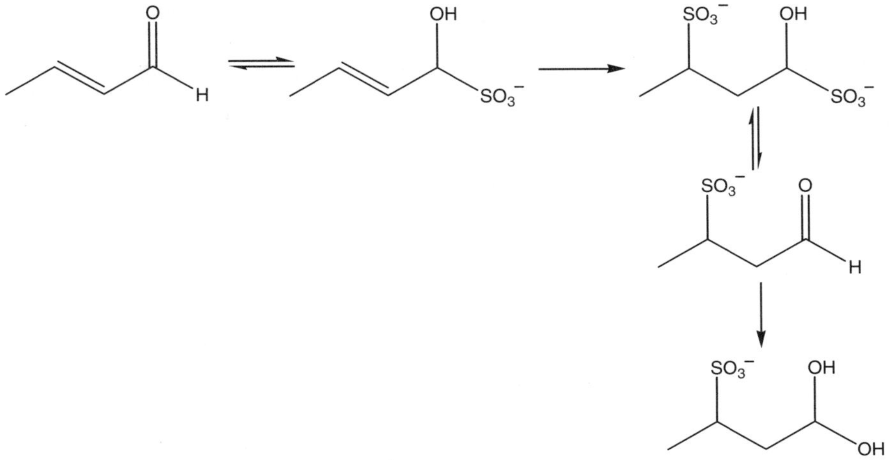

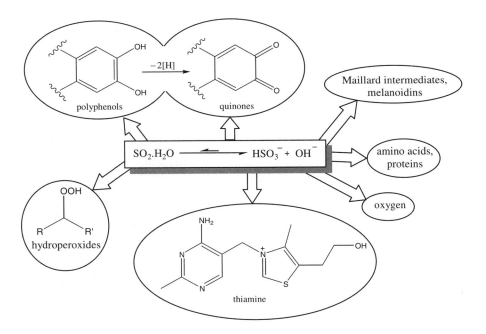

Figure 4.7 *The interaction of sulfur dioxide with α,β-unsaturated aldehydes. There is evidence for the formation of both the reversible bisulphite adduct and the irreversible addition of bisulphite to the carbon–carbon double bond*

Figure 4.8 *Possible routes for the loss of sulfur dioxide in beer. These reactions have been most intensively studied in food systems*

Spin concentration

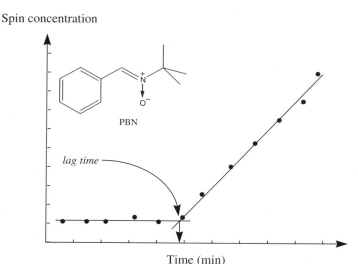

Time (min)

Figure 4.9   *Representation of the derivation of lag times by ESR spectroscopy. The longer the lag time is, the greater the ability the sample has to bind radicals as they are formed and the greater the flavour stability of the beer*

Analytically the changes in bitterness and diacetyl do indeed bear out these sensory observations and have implications for the holistic perception of beer. In addition to this, the sensorially significant esters present in fresh beer can also change, as esters are in equilibrium with their component alcohols and acids. If this equilibrium does not prevail in beer at packaging then, at beer pH, a slow equilibration process can take place. Thus losses in the fruity ethyl esters of hexanoic acid and octanoic acid have been observed for beers in cold storage. The significant levels of nicotinic acid extracted from malt during the brewing operation are also slowly converted to ethyl nicotinate in the presence of ethanol.

Solving Flavour Instability of Beer

There are a number of practical measures which can prolong the sensory integrity of beer during storage. These can be carried out during production, especially with regard to controlling oxygen levels downstream protect the antioxidant portfolio naturally present in beer. In-pack, oxygen scavenging caps, protection from elevated temperatures and efficient distribution chains can all help to preserve beer flavour. However as yet it is difficult to devise a solution for the changes in ester levels in beer during storage, given the relatively vast excess of ethanol present.

FOAM STABILITY

A detailed consideration of beer foam was presented in Chapter 2. Here the factors which can compromise beer foam stability during ageing are described. These issues should not be underestimated: beer foams which do not meet consumer expectations often result in complaints to the brewer, regardless of whether the problem originates from production or at point-of-sale. There are two fundamental issues which compromise beer foam quality. The first is a lack of foam-positive material in the beer to support a stable foam structure, whilst the second is the presence of foam-negative components which reduce foam stability even in the presence of adequate foam-stabilising components.

Lack of sufficient foam-positive components in beer is rare. There are usually sufficient foam-active polypeptides and hop bitter acids in beer to provide a stable foam backbone. If the dispense operation is performed reliably, there is ample gas (carbon dioxide, nitrogen or both) and shear to enable foam to be formed. However, the presence of active proteolytic enzymes in beer can be a source of foam deterioration with time. Proteases can be derived from yeast, either by secretion or cell autolysis, or from enzymic additions made to break down haze-forming proteins post-fermentation. There are a number of protease products available commercially, but the most common are based on papain, derived from the tropical papaya fruit. Normally, such protease activities are reduced or eliminated during pasteurisation, but it is becoming increasingly common for brewing companies to seek alternatives to pasteurisation[a] by a combination of filtration and sterile filling of the package. As no heating is employed here, any residual protease activity can eventually lead to a poorer foam performance.[9] Again, it is worth emphasising that this activity need not be great as the beer in-pack may be stored for weeks or even several months prior to consumption. Other losses may be significant, *e.g.* the loss of iso-α-acids during storage, but this has still yet to be proven.

Losses of foam-active material can potentially compromise final beer foam stability. If such losses occur in-pack though, the major quality issue is likely to be the formation of haze. This is because foam formed in the package does not completely redissolve into the final beer and leaves behind not only haze material in its own right but also nucleation sites for the growth of these nascent particles.

---

[a] There is a global trend to move away from pasteurisation as it is thought by many to reduce the freshness of beers and therefore decrease its quality. Opinion is divided about whether pasteurisation affects flavour stability.

THE FORMATION OF HAZES

Given that beer is rich in polyphenols and polypeptides, it is perhaps unsurprising that beer is prone to the formation of particles with time. With the exception of certain niche beer types, such as wheat beers or agitated bottle fermented beers, this formation of haze is generally unacceptable to the consumer and is a further point of concern for the brewer. The common forms of haze are described below.

Polyphenol–Polypeptide Hazes

Reactions between polyphenols and polypeptides are widespread in society – the traditional process of leather tanning is the irreversible reaction of hide proteins with added polyphenols. Polyphenolic compounds are often termed astringent. This is due to the reaction of polyphenols with salivary proteins, which are predominantly the so-called proline-rich proteins.

Such hazes are the most widespread sources of haze in the brewing industry.[10] The interaction is initially an intermolecular, presumably hydrogen-bonded, complex between the polyphenol and the protein. Further oxidative processes can then result in covalent linkages between the protein and polyphenol moieties. The formation of the initial complex is reversible, whilst the covalently linked aggregates are likely to persist. There is scope for secondary hydrophobic interactions between polypeptides and polyphenols. Many commercial beers can develop so-called chill haze[b] when kept cool for even for relatively short periods of time. This disappears (redissolves) when the beer is allowed to warm up. If the beer is stored for longer, the formation of haze becomes increasingly irreversible, giving rise to permanent haze.[c] The presence of chill haze potentiates the formation of permanent haze, so whilst the formation of chill haze is reversible, its propensity to form should be minimised to ensure good physical stability.

The structure of haze formed in beer has received some attention. The best-known hypothesis is that put forward by Siebert and co-workers.[11,12] The model is based on the idea that only proline-rich proteins can interact with polyphenols to form chill haze and that there is a fixed number of binding sites, implying a more or less definite stoichiometry for haze formation. For the formation of large aggregates, the model also needs to account for cross-linking, so the polyphenols must have at least two protein binding sites (Figure 4.10).

[b] Defined as haze which is present at 0 °C but not at 20 °C.
[c] Defined as haze which is present at 20 °C.

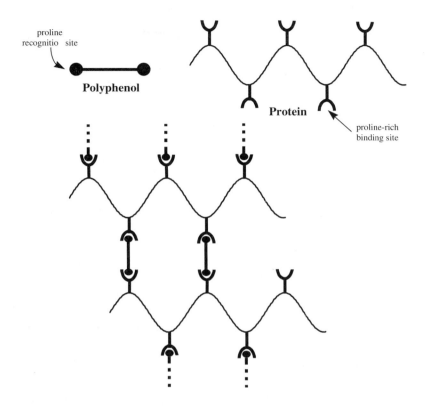

Figure 4.10 *Model proposed by Siebert* et al.[14,15] *for polyphenol–protein interactions in beer. This model implies a more or less definite stoichiometry of cross-linking*

Various downstream operations have been devised to minimise or prevent this from occurring. Evidently, removal of one or the other of the reactive species is one solution. Indeed, there are commercially available silicas which can adsorb beer polypeptides in an effort to reduce their final levels in beer. Many of these products are designed specifically to remove haze-forming polypeptides, as there is a concern that treatment with excess quantities of silica could compromise beer foam stability. However, it is fair to state that there has been no unequivocal proof that foam polypeptides are different from haze polypeptides. Another approach is to remove the polyphenols present. Here again initial research showed that nylon polymers could do this effectively. Today, polymers of vinylpyrrolidone (PVP and its cross-linked variant, PVPP; Figure 4.11) are used which can effectively reduce the polyphenols content of beer typically by a factor of two to three. The key feature of both nylon and

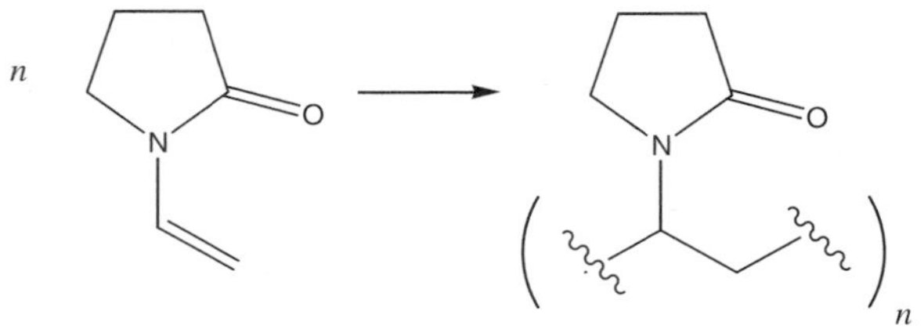

1-vinyl-2-pyrrolidone   polyvinylpyrrolidone (PVP)

Figure 4.11   *Polyvinylpyrrolidone (PVP) and its cross-linked variant (polyvinylpolypyr-*
rolidone, PVPP) are used extensively today in the brewing industry to reduce
polyphenol levels in the final beer. This may either be 'single-shot' or regener-
able with sodium hydroxide solution

PVP/PVPP is the presence of amide linkages so that they retain the hydrogen bonding sites which polyphenols favour on polypeptides.

As with polypeptide removal, the removal of polyphenols should be done judiciously as these compounds may provide protection against the deterioration of beer flavour. Often a combination of treatments is employed in the expectation of finding a compromise solution. The model of polypeptide–polyphenol haze is, though, a little simplistic as metal ions have been shown to accumulate in haze particles and recent reports show that sugars are also present.

Finally, cold filtration is often effective for enhancing the physical shelf-life of beer. Beers freeze at sub-zero temperatures, the exact temperature being dependent on the rate of cooling and the specific beer composition. Ideally, filtration should be carried out close to the freezing point of a particular beer, especially if the beer has been stored cold so that any chill haze has the opportunity to form. Thus removal of as much chill haze as possible prior to packaging has been observed to retard the formation of polypeptide–polyphenol hazes. However, in practical terms, it is important to minimise the risk of beer freezing on the filter, which limits how cold the beer can be filtered.

Calcium Oxalate

Oxalic acid is released from malt during mashing. It forms sparingly soluble salts with calcium,[d] the major divalent cation in beer. If there is insufficient calcium upstream, the oxalate is not effectively precipitated and is a potential 'time bomb' in-pack. Such crystals do not only cause haze but can also provide a multitude of nucleation sites for the nucleation of dissolved gases in beer. On opening the package, the rapid

---

[d] Solubility product of calcium oxalate $= 2.3 \times 10^{-9}\,\mathrm{mol^2\,dm^{-6}}$ in aqueous solution at 298 K.

gas nucleation results in uncontrolled foaming or 'gushing' and, in such situations, can result in substantial beer loss and, in extreme cases, significant dry-cleaning bills!

Carbohydrates

Starch (α-glucan) itself should be degraded upstream by the amylolytic activity during mashing. Any starch remaining, though gelatinised, undergoes a process known as retrogradation when cooled. Retrograded starch is more resistant to enzymic breakdown than conventional starch and, when formed, gives a haze. This haze may be clearly visible or, alternatively might be deemed 'invisible'. The latter, whilst not immediately apparent to the eye, nonetheless scatter light in conventional haze measurements. Similarly, β-glucans, which should also be substantially absent from the final beer, can also give rise to visible and invisible hazes. Wheat is used as an adjunct for the production of certain beers, and is relatively rich in pentosan sugars. Again, such biopolymers can give haze problems in the final product.

Other Sources of Haze in Beer

Some other phenomena can result in haze in the final beer, although they are generally rare. The exogenous foam stabiliser propylene glycol alginate (PGA) can give rise to haze problems under two circumstances. Firstly it should not be added downstream simultaneously with the chillproofing enzyme papain as they tend to react together forming haze. Secondly, PGA can react with beer proteins if the beer experiences elevated temperatures ($> 45\,°C$) for a few days. The latter might seem inconceivable, but exported beers can be exposed to such temperatures during distribution. Dead bacteria from malt, carbohydrates and proteins from damaged yeast and lubricants used in canning or on packaging conveyors can also give haze in the final beer.

MICROBIOLOGICAL CONTAMINATION AND BEER QUALITY

The presence of microbiological contaminants during beer production and in-pack not only compromises beer flavour, but also generates haze as the number of bacteria increase. Flavour defects include an increase in acidity, the formation of volatile sulfur compounds, acetaldehyde, phenolics, diacetyl, ethanol and nitrite. The formation of ethanol is also considered to be a product safety issue, particularly in low- or non-alcoholic beers, whilst nitrite is a known hazard.

Whilst not dangerous, clearly a hazy beer with significant off-flavours is to be avoided. The occurrence of microbiological problems is minimised by continual monitoring, good hygiene, effective cleaning and disinfection, and adequate sterilisation methodology (*e.g.* pasteurisation, sterile filtration). For a comprehensive overview see Priest and Campbell.[13]

Brewery Spoilage Organisms

These may be classified into four groups: Gram-positive bacteria, Gram-negative bacteria, wild (*i.e.* not deliberately introduced) yeast, and moulds. The effects of the most significant microorganisms in terms of beer quality will be discussed in turn.

Gram-positive Bacteria

These are also termed the lactic acid bacteria (or 'lactics') and include the two genera *Pediococcus* spp. and *Lactobacillus* spp. They are responsible for the greatest economic loss and are the most feared in the brewery. Of these, those that are both pH and hop acid resistant spoil beer in-pack by the formation of off-flavours (especially lactic acid and diacetyl), turbidity, rope[e] and some strains can form biogenic amines. These latter are often considered to contribute to the 'morning after' effects of enthusiastic consumption. Although anaerobic, they are aerotolerant and, as such, can occur throughout the brewery. The most significant lactics which impact on beer production are given in Table 4.3.

Some species of *Bacillus* are also Gram-positive. Whilst they cannot spoil beer itself, they are thermophilic, thriving at 55–70 °C. They are capable of producing large quantities of lactic acid both in the mash and in sweet wort and can even be a potential hazard, being able to reduce nitrate to nitrite.

Gram-negative Bacteria

These are generally insensitive to the antibacterial effects of the hop bitter acids. The Gram-negative bacteria are commonly divided into four subdivisions: *Enterobacteriaceae*, acetic acid bacteria, *Zymomonas* spp. and the strict anaerobes. The *Enterobacteriaceae* or wort bacteria are essentially *Obesumbacterium proteus* and *Enterobacter agglomerans*. They

---

[e] Rope is a complex polymeric material which gives beer an unpleasant thick, stringy texture.

Table 4.3 *Significant lactic beer spoilers*

Lactic acid bacterium	Metabolism*	Comments
Lactobacillus		
L. brevis	heterofermentative	Can convert pentoses and maltose to lactate
L. casei	homofermentative	
L. plantarum	homofermentative	
L. delbrueckii	homofermentative	Thermophilic
Pediococcus		
P. damnosus	homofermentative	*Pediococcus* can continue to grow in deposited yeast after the cessation of fermentation
P. pentosaceus	homofermentative	

*Homofermentative lactics convert a high proportion of sugars present into lactate. Heterofermentative lactics also produce acetate, ethanol and CO_2 in addition to lactate.

grow in wort at the early stages of fermentation, reducing the rate of fermentation, generating dimethyl sulfide and also reducing nitrate. The acetic acid bacteria (*Acetobacter* spp., *Gluconobacter* spp.) are aerobic and can be problematic in cask-conditioned beer and dispense equipment. They can proliferate rapidly, forming acetic acid and giving the beer a vinegar-like flavour. In addition they can lead to haze and rope. *Zymomonas* is a bacterium which has historically led to the closure of breweries when it has been persistent. It cannot metabolise maltose, but readily grows on glucose and fructose, components of priming sugars used for cask ale production. Its main impact on cask-conditioned beer is the production of acetaldehyde and hydrogen sulfide, a combination which is reminiscent of rotting apples. It also leads to turbidity.

More recently, the strict anaerobes *Pectinatus* and *Megasphaera* have created beer spoilage incidents in a number of breweries worldwide. They are intolerant even of low oxygen levels, and are most often present in bottled beers. They produce a range of organic acids (particularly acetic acid and, in the case of *Pectinatus*, propionic acid), leading to increased acidity, as well as hydrogen sulphide. *Pectinatus*, though, is recognised as being the most prolific inducer of haze of all the brewery bacteria. Ironically, the move towards ever tighter oxygen control in-pack provides just the right environment for these microbial contaminants to thrive.

Wild Yeasts

The *Saccharomyces* wild yeasts, exemplified by *Saccharomyces cerevisiae* var. *diastaticus* spoil beer by the formation of haze. In addition they tend to overattenuate the beer, metabolising the residual sugars to give more alcohol and carbon dioxide, so that the final beer is out of specification for these two parameters. They are also able to decarboxylate ferulic acid to give rise to the highly characteristic 4-vinylguaiacol (see Chapter 3). The non-*Saccharomyces* wild yeasts (*Brettanomyces*, *Kloercka*, *Pichia*, *Candida* and *Hansenula*) produce a range of off-flavours, haze, overattenuate and affect the fermentation pathway. *Brettanomyces* also forms organic acids. Whilst these wild yeasts are found throughout the brewery, they tend to be most problematic in cask-conditioned beers.

Moulds

Moulds are rarely a problem unless oxygen is available. However, poorly cleaned equipment contaminated with residual beer can turn mouldy as nutrients, air and oxygen are present. Moulds can also develop in inadequately sealed bottles, growing at the air–beer interface. The use of air pressure to dispense beer can also give rise to moulds in dispensing equipment. Moulds, though, make their greatest impact on barley and during malt production. Growth of these moulds can lead to the formation of harmful toxins, such as aflatoxins and ochratoxin A. The filamentous fungus *Fusarium* can grow on barley to produce compounds which are potent gushing promoters.

SUMMARY

The complexity of beer composition and production provides numerous options for its deterioration over time. Of particular note is the degradation of flavour by routes which though strongly suspected, have not been unequivocally proven. Similarly, blame for the reduction on foam quality with time is apportioned to the activity of proteolytic enzymes in the beer and in this context is likely to increase in importance as breweries increasingly rely on non-thermal methods of ensuring the microbial integrity of beer. Microbiological contamination is ever-present and control can only be maintained by being vigilant, not only during production, but by rapid and sensitive testing of the final product.

REFERENCES

1 C.W. Bamforth, R.E. Muller and M.D. Walker, 'Oxygen and oxygen radicals in malting and brewing: A review', *J. Am. Soc. Brew. Chem.*, 1993, **51**, 79–88.
2 T. Shimazu, N. Hashimoto and T. Eshima, 'Oxidative degradation of

isohumulones in relation to hoppy aroma of beer', *Rep. Res. Lab. Kirin Brew. Co.*, 1978, **21**, 15–26.

3 N. Hashimoto, T. Shimazu and T. Eshima, 'Oxidative degradation of isohumulones in relation to beer flavor', *Rep. Res. Lab. Kirin Brew. Co.*, 1979, **22**, 1–10.

4 M.C. Meilgaard, 'Flavor chemistry of beer: Part II: Flavor and threshold of 239 aroma volatiles', *Master Brew. Assoc. Am. Tech. Q.*, 1975, **12**, 151–168.

5 P.A.T. Swarboda and K.E. Peers, 'Metallic odour caused by vinyl ketones formed in the oxidation of butterfat. The identification of octa-1,*cis*-5-dien-3-one', *J. Sci. Food Agric.*, 1977, **28**, 1019–1024.

6 S. Sakuma and M. Kowaka, 'Flavor characteristics of *cis*-3-nonenal in beer', *J. Am. Soc. Brew. Chem.*, 1994, **52**, 37–41.

7 M. Nyborg, H. Outtrup and T. Dreyer, 'Investigation of the protective mechanism of sulfite against beer staling and formation of adducts with *trans*-2-nonenal', *J. Am. Soc. Brew. Chem.*, 1999, **57**, 24–28.

8 J-P. Dufour, M. Leus, A.J. Baxter and A.R. Hayman, 'Characterization of the reaction of bisulfite with unsaturated aldehydes in a beer model system using nuclear magnetic resonance spectroscopy', *J. Am. Soc. Brew. Chem.*, 1999, **57**, 138–144.

9 D.R. Ilett, 'Aspects of the analysis, role, and fate of sulfur dioxide in beer – A review', *Master Brew. Assoc. Am. Tech. Q.*, 1995, **32**, 213–221.

10 C. Forster, J. Schweiger, L. Narziss, W. Back, M. Uchida, M. Ono and K. Yamagi, 'Investigation into the flavour stability of beer by means of electron spin resonance spectroscopy of free radicals', *Monatsschrift Brauwissenschaft*, 1999 (May/June), **52 (5/6)**, 86–93.

11 C.E. Dalgleish, 'Flavour stability', *Proc. Eur. Brew. Conv. Cong.*, 1977, 623–659.

12 M. Muldbjerg, M. Meldal, K. Breddam and P. Sigsgaard, 'Protease activity and correlation to foam', *Proc. Eur. Brew. Conv. Cong.*, 1993, 357–364.

13 C.W. Bamforth, 'Beer haze', *J. Am. Soc. Brew. Chem.*, 1999, **57**, 81–90.

14 K.J. Siebert, N.V. Troukhanova and P.Y. Lynn, 'Nature of polyphenol protein interactions', *J. Agric. Food Chem.*, 1996, **44**, 80–85.

15 K.J. Siebert and P.Y. Lynn, 'Comparison of polyphenol interactions with polyvinylpolypyrrolidone and haze-active protein', *J. Am. Soc. Brew. Chem.*, 1998, **56**, 24–31.

16 F.G. Priest and I. Campbell, *Brewing Microbiology*, Chapman and Hall, London, 1996.

Chapter 5

Nutritional Aspects of Beer

BEER COMPONENTS OF NUTRITIONAL VALUE

It has been recognised for a very long time that beer can make an important contribution to the diet. For early man, the brewing of beer was closely related to the baking of bread since both were made from grain, water and yeast, and both were broadly similar in nutritional value. By the time the ancient cities of Mesopotamia were built 6000 years before Christ, the brewing of beer was well established. These beers would probably have been quite thick, higher in protein, lower in alcohol, and sweeter than those we are accustomed to today. The Egyptians also made beer, and taught the art to the Greeks and Romans, beer being particularly popular in the Roman empire. Gradually however, wine became the accustomed drink of the Mediterranean countries while beer migrated north. Throughout the medieval period beer was drunk in large quantities by most of the population, since it was microbiologically safer than water, partly because it was boiled and partly because of the anti-bacteriological properties of alcohol and, after they were introduced in the 15th century, hops. Ladies of the Elizabethan court received an allowance of two gallons of beer a day! Even the children drank the lower alcohol 'small beer'.

Sorghum beers still provide a significant proportion of the dietary protein and vitamins for much of the poorer areas in sub-Saharan Africa. However, for Western affluent societies, the protein contribution is perhaps less important than it used to be. The main constituents of modern beer are listed in Table 5.1 and compared with milk, wine and carbonated soft drinks. It is apparent that beer contains significantly more of the main nutrients (proteins, carbohydrates, fibre and vitamins) than other alcoholic beverages, such as wine and non-alcoholic beverages such as lemonade or colas. Indeed, its vitamin content is easily as good as that of

Table 5.1 *Typical beer composition – major constituents*

Ingredient	Beer	Wine	Milk	Carbonated soft drinks
	Typical level (g/100 ml)			
Water	92–95	85–91	88–90	89
Alcohol	2.5–3.5	9–14	0	0
Total carbohydrates	1.5–3	0.1–6.0	5	10
of which free sugars	<0.2	0.1–6.0		10
Total proteins, peptides and amino acids	0.2–0.6	0.02	3	negligible
Lipids	negligible	negligible	3–4	0
Minerals	0.2–0.3	0.1–0.3	0.2–0.5	0.025
Vitamins and other micronutrients	0.002	0.0003	0.002	0
Fibre	0.3–1.0	negligible	negligible	negligible
Polyphenols and hop compounds	0.002–0.06	0.03–0.074	0	0

Sources: McCance and Widdowson's *Composition of Foods*, 4th edition, HMSO, London, 1978.
D. E Briggs, J. S. Hough, R. Stevens and T. W. Young, *Malting and Brewing Science*, Volume 2, Chapman & Hall, London, reprinted 1986.

milk, while the fat content is much lower. Milk does, however, contain more protein than beer.

Each of these nutrient categories will be discussed separately.

Water

A sufficient intake of fluid is important for maintaining electrolyte balance between the inter- and intra-cellular fluids in the body, and to counteract dehydration. A daily intake of 2.5 litres of water is normally recommended for an adult male in temperate climates, and more in the tropics or with increased levels of exercise, such as physical labour or participation in active sports. Beer has long been recognised as a palatable way of providing this essential fluid. Indeed, the British style of beer known as 'Mild' – which is relatively low in alcohol (around 3% ABV) and not too bitter – largely developed in areas such as Yorkshire, Northumberland and Nottinghamshire where there was substantial heavy industry in the steelworks, coal-mines and shipbuilding yards. A modern equivalent of these beers are the low alcohol lagers and alcohol-free malt liquors developed and promoted in Germany as sports drinks. These can rapidly replace the fluid and minerals lost in sweat as well as providing readily available calories to restore energy levels.[1] Some clinical studies

Table 5.2 *UK beer market – alcohol content*

Alcohol content (% by volume)	% of sales volume
< 1.1 (No- and low-alcohol)	0.2
1.3–4.2	63.0
4.2–7.5	23.4
> 7.5	2.6
Not classified according to alcohol content (includes stouts and non-returnable packages)	10.7

Source: BLRA Statistical Handbook, 1998 edition.

have also shown that beer is more effective than water in flushing out the kidneys and offers some protection against kidney stones,[2,3] which are a particular problem for runners in hot climates.[1] The ingredients of beer which are most active in this respect are not proven, but are most likely the mineral ions, which are examined in more detail later in this chapter.

Alcohol

The alcohol content of beer is expressed in many countries, including the UK, as % ABV (% alcohol by volume). In others such as the USA it is expressed by weight, which gives slightly lower figures (4.8% alcohol by volume would be around 3.8% by weight). Across the world, the alcohol content in beers ranges from less than 0.05% ABV in alcohol-free beers to 12.5% ABV in the British Thomas Hardy Ale, which is one of the world's strongest beers. However, the majority of beers sold in the UK are below 4.2% ABV (Table 5.2). In many countries, the rate of excise duty is proportionally higher at higher alcohol contents, so that the majority of brands fall into the moderate bracket. (In Mexico, for example, beverages with less than 5% by volume of alcohol are classified as non-intoxicating!) These moderate levels of alcohol – on average a half of the level in wine – mean that the beer drinker can consume quite large volumes of fluid without excessive alcohol intake. In the UK the sensible drinking guidelines recommended by the Department of Health are 3–4 units a day (about 2 pints of standard strength beer) for men and 2–3 units for women. Units of alcohol are explained in more detail later in this chapter (see 'Risks and Benefits of Drinking Alcohol').

It is now becoming widely recognised that moderate consumption of alcohol can confer distinct health benefits. A large number of studies have compared mortality in populations with their intake of alcoholic beverages. Most of these studies show that there are fewer deaths from all causes amongst people who drink one or two drinks a day compared with people who abstain from alcohol completely or who consume it only

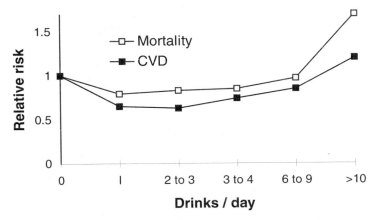

Figure 5.1   *Alcohol consumption, mortality from all causes and deaths from cardiovascular disease*

rarely.[4–8] Figure 5.1 shows average findings from a number of these studies, expressed in terms of relative risk in relation to alcohol intake. At intakes in excess of around 5–9 drinks a day, deaths from alcohol-enhanced causes (such as accidents and cirrhosis of the liver) start to exceed the protective effects. The relative risks and benefits of alcohol consumption are briefly discussed under the appropriate section later in this chapter.

A large part of the protective effect is due to a reduction in deaths from cardiovascular diseases such as heart attacks. These findings are supported by a very large number of trials in many countries, covering men and women from different ethnic groups and different cultures, including European, North American, Chinese, Australian and Japanese. These studies have recently been reviewed by Chick.[9] The mechanisms by which alcohol might exert these protective effects have been explored in a number of *in vivo* and *in vitro* studies. These indicate that alcohol consumption is associated with an increase in high density lipoproteins (HDL_2 and HDL_3) and apolipoproteins A-I and A-II which are considered to be protective against coronary heart disease[10–12] and a decrease in the low density lipoproteins which are associated with atherosclerosis. In addition, alcohol appears to reduce the likelihood of blood clotting by lowering the concentration of fibrinogen (a blood clotting agent) in the blood plasma[13] and also by reducing the tendency of blood platelets to aggregate.[14]

Table 5.3 *Carbohydrates in beer*

Beer type	Sugars (including maltotriose)	Higher dextrins	Total
Ales and stouts	0.5–3.0	1–4	1.5–6
Primed beers	1.3–3.6	1–4	2–7
Lagers	0.1–0.7	1–2	1–3
'Lite' beers	0.1–0.6	0.1–0.3	0.2–0.9

Carbohydrate (g/100 ml) is the spanning header over the Sugars, Higher dextrins and Total columns.

Carbohydrates

Levels of total carbohydrate in beer range from 2 to 3 g/100 ml (see Table 5.3), which are comparable to those in milk, where the carbohydrate content is typically around 5 g/100 ml. The source of the fermentable sugars in beer is starch-rich cereals, mainly malted barley (see Chapter 1). Other cereals, both malted and unmalted, including wheat, rice, maize, oats and sorghum, are used for certain beers and sugar syrups may also be used as an additional source of fermentable sugars. The major effect of the malting process is to break down some of the cellular materials (such as cell walls and storage proteins) which can cause difficulties in processing and to develop the hydrolytic enzymes, particularly amylases, which are needed to digest the starch and convert it to fermentable sugars. This digestion of starch takes place during the mashing stage of brewing, not during malting. Therefore, provided that there is sufficient malt in the mash to provide adequate amylolytic activity, both malted and unmalted cereals provide a similar range of fermentable sugars and unfermentable, mainly branched, dextrins (see Chapter 1) in the wort.

During fermentation the yeast absorbs and ferments first all the glucose and then maltose. Some yeasts, particularly lager yeasts, can also utilise maltotriose, which contains three glucose units. Thus lagers are in general more fully fermented (attenuated) than ales, and contain less residual carbohydrate. However the higher dextrins, both branched and straight-chain cannot be fermented by the yeast. Thus beer contains little in the way of free sugars, but a variable range of higher dextrins, most of which contain two or more branches. These dextrins have little effect on sweetness but probably do contribute to the mouthfeel and the perception of body in the beer, especially in British ales. Some traditional British beers, such as some ales and sweet stouts, may contain additional sugars (primings) which are added after fermentation to give a sweeter product or sometimes to balance excessive bitterness from high levels of hops. Low carbohydrate beers, on the other hand, contain less carbohydrate because the dextrins have been more or less completely digested and

Table 5.4 *A comparison of protein in some common beverages*

Beverage	Protein $(g\,l^{-1})$
Beer	2–7
Wine	1–7
Carbonated soft drinks	negligible
Spirits	negligible
Milk	32–35

Source: McCance and Widdowson's *The Composition of Foods*, 4th edition 1978, HMSO, London.

fermented by a combination of techniques. For example, lightly kilned malts, which contain higher levels of debranching enzyme, and special yeasts (known as 'highly attenuating yeasts') which can digest a wider range of dextrins, may be used.

When beer is drunk it is unlikely to remain in the mouth long enough to be acted upon by salivary amylase, although this enzyme is capable of digesting straight chain dextrins. Such dextrins are readily hydrolysed by pancreatic amylase, which is an α-amylase. The branched dextrins are broken down more slowly by the oligo-1,6-glucosidase which is found in the mucosal cells lining the intestine. Thus the rise in blood sugar is slower after drinking beer than after consuming a soft drink with a similar loading of carbohydrate, since in the latter the carbohydrate is largely in the form of simple sugars.

Proteins, Peptides and Amino Acids

Beer contains on average between 0.2 and 0.6 g/100 ml of protein-derived material, which is substantially more than other alcoholic beverages such as wine,[15,16] although less than high protein beverages such as milk (see Table 5.4).

The origin of this protein is malted barley, which contains around 10–12% protein, about a third of which is extracted during mashing. Many of the larger proteins (that is, with a molecular weight above about 17 kD), are precipitated out and removed during the wort boiling stage, while most of the free amino acids in wort are taken up by the yeast during fermentation. Thus most of the protein-derived material in beer is in the form of peptides and polypeptides, which are readily digested. The imino acid proline cannot be assimilated by yeast and this represents a substantial proportion of the low molecular weight nitrogenous material in beer[17] (see Table 5.5). Nevertheless, most beers contain all the essential amino acids, at levels generally between 5 and 10 mg per 100 g.[18]

Table 5.5 *Amino acids, peptides and proteins in beer*

Size range	% of total nitrogenous material
Amino acids and proline	5–11
Small peptides (< 5000 MW)	50–60
Large peptides (> 5000 MW)	25–45

Lipids

Malted barley contains only around 3% of lipid, mainly in the form of triglycerides. These are not readily extracted into aqueous solution, but limited lipase activity during malting and mashing releases some fatty acids, mainly unsaturated C_{18} linoleic and linolenic acids, into the wort. Most of these are taken up by the yeast and used for yeast growth. Beer is thus a very low fat food, containing only traces of lipid. The low levels of lipid are also important for beer quality, since they can have deleterious effects on both foam and beer flavour (see Chapters 2 and 3). Of course, both alcohol and carbohydrate can be converted to fat in the body, if consumption exceeds calorific requirements.

Fibre

Surprising as it may seem, beer does contain fibre! Dietary fibre is generally defined as non-starch polysaccharides, and in beer this comes from the β-glucan cell walls in barley. The high molecular weight β-glucan can cause separation and filtration problems during brewing, so should be degraded during malting (see Chapter 1). However, if it is sufficiently degraded the smaller, soluble glucans do not cause processing problems and can make a useful contribution to dietary fibre, since they still contain the all-important β-linkage. On average beer contains between 0.3 and 1 g/100 ml of fibre (Table 5.1). In addition to contributing to the healthy functioning of the large intestine, the β-glucans in fibre can lower the levels of cholesterol in the blood serum,[19] thus protecting against coronary heart disease.

Energy Value

The calorific (energy) value of most beers is between 20 and 40 kcal/100 ml (80 to 180 kJ/100 ml) (see Table 5.6). This is similar to milk (typically around 200 kJl/100 ml). In beer, much of the energy is derived from the alcohol, together with the residual carbohydrate and protein, whereas in milk there is a greater contribution from protein and fat.

Table 5.6 *Typical energy values of some beer types*

Beer type	Energy value	
	kJ/100 ml	*kcals/100 ml*
'Lite' beers	72–110	20–26
Lagers	85–125	20–30
Ales	114–160	25–38
Stouts/strong beers	150–300	35–70

In the UK the energy value of beer is calculated from the content of alcohol, carbohydrate and protein according to a formula defined by the Food Labelling Regulations:

$$\text{Energy value (kJ/100 ml)} = (\text{alcohol} \times 29) + (\text{carbohydrate} \times 17) + (\text{protein} \times 17) \qquad (5.1)$$

where alcohol, carbohydrate and protein are all in g/100 ml.

It is evident, therefore, that the alcohol content makes the largest single contribution to the energy value. Thus low carbohydrate beers can still have a relatively high energy value if the alcohol content is significant. This can lead to some confusion since the term 'lite' is variously applied to low calorie, low carbohydrate and low alcohol beers in different countries.

Minerals

Beer contains a wide range of mineral ions, as would be expected in a beverage which is mainly water. Since the composition of water varies widely both within and between countries, the mineral content of beers can vary likewise (see Table 5.7). Indeed, the different styles of beer which have developed across the world depend to a considerable extent on the composition of the natural water supply, since mineral ions can have a significant effect on flavour. Thus the calcium- and sulfate-rich waters of much of England, particularly Burton-on-Trent, are associated with Pale Ales, while the low mineralisation of water in Pilsen in the Czech republic has given rise to Pils-type lagers.

Nowadays it is common to adjust the composition of the brewing water so that the same beer can be manufactured at several different sites, often in different countries. One particular brand or type of beer will thus display a broadly similar mineral composition wherever it is made, but can differ sharply from another beer type made at the same site. However,

Table 5.7 *Comparison of the mineral content of drinking water and beer in the UK*

Mineral	Water ($mg\,l^{-1}$)	Beer ($mg\,l^{-1}$) Range	Beer ($mg\,l^{-1}$) Average	Reference intake for adult male ($mg\,day^{-1}$)[a]	% of Reference intake supplied by 1 pint beer
Potassium	5–10	100–700	450	3500	6
Magnesium	10–50	40–200	70	300	12
Calcium	20–400	40–250	120	700	9
Zinc	5 (max)	0.01–0.17	0.035	9.5	0.2
Sodium	20–110	10–130	80	1600	2.5
Silicon	0.2–8	30–80	—	—	—
Phosphorus	2.2 (max)	90–400	200	550	18
Sulfate	10–600	100–700	120	—	—
Chloride	10–70	100–500	300	2500	6

[a]British Nutrition Foundation, 1992.

the concentration of mineral ions in the beer is not always the same as in the brewing water. Some ions, particularly those of heavy metals, tend to be adsorbed onto the spent grains or the yeast and are removed. Thus beer often contains less of these ions than the water used. Other ions, such as calcium, are removed by precipitation as carbonates and phytates during wort boiling. Other ions may be extracted from the malt or the hops, and are thus more concentrated in the beer than in the brewing water.

In health terms, the most important minerals in beer are potassium, calcium, magnesium and phosphorus and one pint of beer can provide on average around 10% of the daily requirements of these essential elements. This is particularly important for potassium and magnesium since a recent survey by the Ministry of Agriculture, Food and Fisheries indicated that the daily intake of these two minerals was on average 10–20% below requirements.[20] Particularly significant in health terms is the low ratio of sodium to potassium in beer, which is desirable in terms of regulating blood pressure. The intake of sodium in Western societies is generally well above recommended levels and the usual advice is to limit intake to not more than 6 g per day.

Beer is also rich in a bioavailable form of silicon. This element has been recognised as essential for some species, and deficiency can affect the synthesis of healthy bone. Silicon is now thought to play an important role in reducing the levels of aluminium in the body.[21] Aluminium is widespread in the environment and is toxic in high concentrations. In the past it has been considered that aluminium was causally linked to Alzheimer's disease[22] but this mechanism is no longer considered to be valid

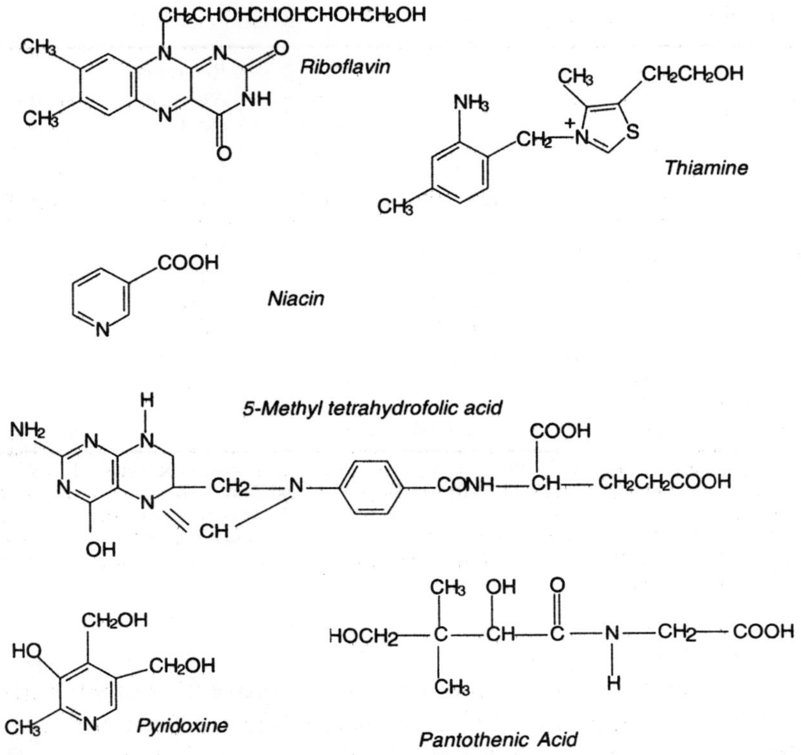

Figure 5.2 *Structural formulae of some important vitamins in beer*

although Alzheimer sufferers may be more at risk from dietary aluminium.[23]

Vitamins and Micronutrients

Beer is a good source of a number of vitamins, particularly many of the B vitamins (see Figure 5.2). These water-soluble vitamins often occur together in natural foodstuffs, and plants and sprouted cereals such as malt are especially good sources. As is evident from Table 5.8, a single pint of beer can provide, on average, 10 to 20% of the daily requirements of niacin, riboflavin, pyridoxine and folate. Thiamine is the exception. Although there are significant quantities in the malt, thiamine is taken up by the yeast so that little is left in the beer. Beer is not a particularly good source of vitamin C, since this vitamin is very heat-labile and is easily destroyed by wort boiling and by pasteurisation. Some bottled beer, however, may contain vitamin C added at up to $30 \, \text{mg} \, l^{-1}$ to protect

Table 5.8 *Vitamin content of beers*

Vitamin	Range in beer[a,b,c] (mg l⁻¹)	Range in wine[f,g] (mg l⁻¹)	Typical level in UK beer[a,c,d] (mg l⁻¹)	Reference Nutrient Intake[e] (RNI) (adult man, mg day⁻¹)	Average % of RNI from one pint of beer
Niacin	3–20	0.8–1.9	7.7	17	20
Riboflavin	0.07–1.3	0.06–0.4	0.3	1.3	12
Pyridoxine (B₆)	0.13–1.7	0.1–0.45	0.5	1.4	18
Foliates	0.03–0.10	0.002	0.05	0.2	12
Biotin	0.007–0.018	negligible	0.01	—	—
B₁₂	0.09–0.14	<0.001	0.1	1.5	3
Pantothenic acid	0.5–2.7	0.5–1.2	0.9	—	—
Thiamine	0.002–0.14	0.005–0.04	0.03	0.9	1.6

Sources: [a]A. Piendl, *Brau. Ind.*, 1982–1985. [b]W. A. Hardwick, in *Handbook of Brewing*, Marcel Dekker, New York, 1994. [c]J. S.Hough, D. E. Briggs, R. Stevens and T. W. Young, in *Malting and Brewing Science, Volume 2*, Chapman and Hall, London, 1982. [d]Brewing Research International, unpublished results. [e]*Dietary Reference Values*, British Nutrition Foundation, 1992. [f]M. Malanda and P. Millet, *Bios*, 1992, **23(6/7)**, 92. [g]McCance and Widdowson's *The Composition of Foods*, 4th edition, HMSO, London, 1978.

against the oxidation of trace lipids and thus improve flavour stability.

Niacin, riboflavin and thiamine are essential for utilisation of carbo-hydrates and fats in the diet while pyridoxine has a similar role in protein metabolism. Folate acts as a carrier for methyl and formyl groups, and as such is essential for DNA synthesis. Evidence is now emerging to suggest that folate may protect against cardiovascular disease and some cancers by regulating levels of homocysteine in blood plasma. Some very recent work suggests that homocysteine levels are lower and vitamin B₆ (which is itself inversely correlated with cardiovascular disease) levels are higher after consumption of beer rather than wine, spirits or water (the con-trol).[24]

Apart from riboflavin, where as much as half can be contributed by the yeast, most of these vitamins are derived from the malted barley, and are thus at higher levels than in alcoholic beverages such as wine, which are not derived from cereals. Beers prepared with significant proportions of unmalted adjuncts, such as rice or maize grits, generally contain less vitamins, while all-malt beers and stouts are usually towards the higher end of the range.

It must be emphasised that the significant contribution beer can make to dietary intake of B vitamins is only relevant for moderate alcohol intakes. Alcoholics are frequently deficient in vitamins, partially because they often have a poor diet, but also because, at high concentrations,

Table 5.9 *Classification of phenolic compounds in beer*

Type of phenol	Example in beer	Typical concentration in beer ($mg\ l^{-1}$)
Monophenols		10–30
phenolic alcohols	tyrosol	
phenolic acids	ferulic acid	
phenolic amines and amino acids	hordenine, tyrosine	
Monomeric Polyphenols		1–23
flavan-3-ols	catechin, epicatechin	
flavan-3,4-diols	leucocyanidin	
flavonols	quercetin	
Condensed Polyphenols		20–140
dimeric and polymeric catechin		
proanthocyanidins	procyanidin B3	
prodelphinidins	delphinidin B3	

alcohol in the diet can interfere both with the absorption and the metabolism of vitamins.

Phenolic Compounds

Beer contains a wide range of phenolic materials, derived both from the hops and the malt. The nomenclature and classification of plant phenolics is confusing, and terms such as flavonoids, tannins and polyphenols are often used rather loosely to cover an ill-defined range of compounds. One simple classification scheme for beer phenolics is given in Table 5.9. In this scheme, a distinction is made between flavanols, such as catechin, and flavonols, such as quercetin, which have a keto group in position 4 (see Figure 5.3).

The proanthocyanidins are closely related structurally to the anthocyanidins, which are the red and blue pigments widely distributed in the plant kingdom, and found, for example, in red wine, tea and coloured fruit and vegetables such as tomatoes. Many of these polyphenols have anti-oxidant properties in laboratory assay systems and there is increasing discussion as to whether such components in the diet can protect against cardiovascular disease by reducing oxidative damage, for example to the blood vessels. Populations which consume a diet rich in fruit and vegetables – often washed down with wine – appear to suffer less from cardiovascular diseases than populations whose intake of anti-oxidants is lower. Some – although not all – epidemiological studies have linked intake of flavanols, sometimes from specific foods such as onions or apples, with reduced risk of coronary heart disease or stroke. The

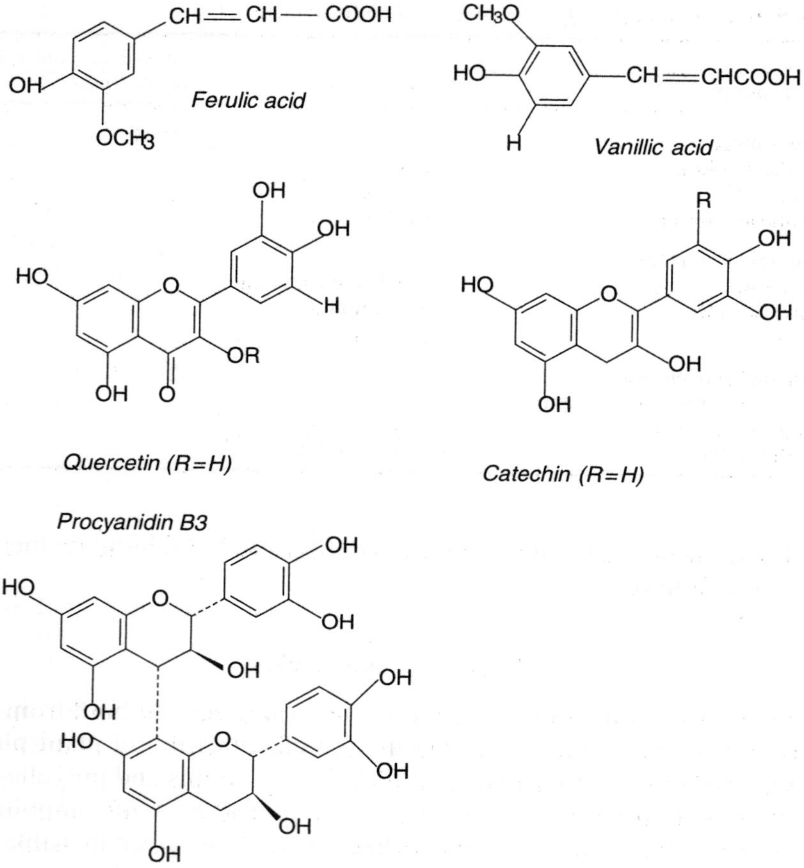

Figure 5.3 *Structural formulae of some phenols in beer*

'French paradox' is often cited – this refers to the lower prevalence of heart attacks in France than in other countries such as the UK, where the consumption of saturated fat is similar, and the difference has been ascribed to red wine in the French diet. Unfortunately this apparently simple picture is confused by a number of interfering factors. Critics of this view suggest that the consumption of saturated fat has increased in France only relatively recently, too soon to have an effect on cardiovascular disease.[25] Other Mediterranean countries such as Greece and Italy continue to eat less saturated fat as well as consuming more vegetables and red wine. Epidemiological studies of alcoholic beverage consumption and cardiovascular disease often cannot differentiate accurately between types of drinks consumed, since most people drink more than one sort of beverage. Where people do consume one type of alcoholic drink in preference to others, there are frequently other variables. For example, in non-wine growing areas, wine is often more expensive, and thus wine drinkers tend to be more affluent and better nourished than the general population. There is also a school of thought which suggests that drinking habits are important; regular consumption with food, which is the way wine is most often consumed, is thought to be more beneficial than drinking without food. Most of the published studies conclude that the protective effects against heart disease are mainly due to the alcohol itself rather than to the type of drink. However, the possibility of additional, smaller effects due to other components in different drinks has not been completely ruled out. Studies by the Free Radical Research Group in Italy suggest that consumption of beer does increase the antioxidant capacity of blood plasma, as well as the content of phenolic acids.[26] In these studies, ethanol alone had no effect on antioxidants, suggesting that this could be a beverage specific effect. However, ethanol was apparently necessary for the efficient absorption of the anti-oxidant phenolics.

Our understanding of the health effects of plant phenolics is still very restricted. Although we can measure anti-oxidant capacity of a certain dietary component in the laboratory, other questions must be answered before we can judge whether it has any effect in the body.

1. How much of the component is present, in which foods, and how much of these foods are normally consumed?
2. Can the component in question be absorbed by the body from the food matrix *i.e.* is it bioavailable?
3. Even if it is absorbed, does it reach vulnerable tissues?
4. Does it exert the same anti-oxidant effect within the body as is observed in the laboratory?

Most of the work to date is concentrating on answering the first question. We are beginning to get some answers to the second question but there is still some way to go. For example, bioavailability is likely to be influenced by the type, and especially the size, of each phenolic component, with the smaller species expected to be more readily absorbed. In beer, a significant proportion of the phenolics are present in the monomeric form as, for example, the hydroxycinnamic acids (such as *p*-coumaric, ferulic, chlorogenic and caffeic acids, together with their esters and gycosides) and the phenolic acids (such as gallic acid and flavon-3-ols) (see Figure 5.3 and Table 5.9). The major hyroxycinnamic acid in beer is ferulic acid, which is derived largely from the barley cell walls. Levels in beer range from 0.52 to 2.36 mg l^{-1} and recent studies have shown that it is readily bioavailable to humans.[27] It is less certain whether the condensed polyphenols, such as the anthocyanins, can be as easily absorbed.

Little information is available to answer questions 3 and 4. It is probable therefore that this debate will continue for some time until more hard evidence is available.

Recently published research suggests that some hop flavonoids can reduce cell proliferation in human breast and colon cell lines,[28] and also inhibit activation of pro-carcinogens within cells.[29] These results are very interesting, but it is as yet too early to make any deductions regarding possible implications for human health, since not enough is known about the effects on whole animals.

Other related flavonoids in hops may have phytoestrogenic characteristics. Phytoestrogens are compounds which occur naturally in some plants and can interact with animal metabolic systems in the same way as the hormone oestrogen, but are very much weaker. They are found in a number of plants, including cereals, with soya beans being a particularly rich source. Epidemiological evidence from countries such as Japan, where soya is part of the staple diet, suggests that high intakes of phytoestrogens may protect against certain cancers, particularly those which are sensitive to hormones, such as prostate and breast cancer, and also against cardiovascular disease.[30] Anecdotal information has for some time suggested that hops could contain phytoestrogens but the chemical identity of these compounds – if any – has been elusive. Recently published work suggests that the active agent may be a prenyl flavonoid.[31] Although hops are relatively rich in this flavonoid, current levels in beer are too low to be of any clinical significance.

Hop Bitter Acids

Beer is unique in its use of hops to provide an essential part of its characteristic flavour and aroma. Hops are thus the one ingredient which distinguishes beer from all other beverages. In health terms, the most important attribute of hops is their anti-microbial effect, which is one of the main reasons why beer is so resistant to bacterial spoilage. The α-acids (the bittering agents in hops) are ionophores which can act on the plasma membrane of cells to break down the trans-membrane proton gradient. This gradient provides the driving force by which cells take up nutrients, thus if it is inactivated the cells effectively starve to death.[32] The Gram-positive group of bacteria, which includes a number of food poisoning organisms such as *Listeria*, *Staphylococcus*, some *Streptococcus*, *Bacillus cereus* and *Clostridia*, are all killed by hop acids and thus cannot survive in beer. The Gram-negative group, which have an outer cell membrane to protect the plasma membrane, are resistant to hop acids, but these bacteria are sensitive to other properties of beer, such as the low pH, the alcohol, or the low oxygen concentration. A small group of organisms, such as some of the *Lactobacilli* and *Acetobacilli*, have evolved which are resistant to hop acids, and these are capable of growing in beer, causing hazes and off-flavours, but these spoilage bacteria do not include any disease-causing organisms.

METABOLISM OF ALCOHOL

Alcohol is readily absorbed from the diet, with most absorption taking place in the small intestine. The rate at which alcohol is absorbed varies widely between individuals, and is affected by a number of external factors relating to the alcoholic beverage itself, the pattern and rate of drinking, other food consumed and physiological factors such as body mass, previous exposure to alcohol and genetic factors. Only a very generalised picture can be given here.

When alcohol is ingested a small amount may be metabolised directly in the stomach by gastric alcohol dehydrogenase.[33] The extent of gastric metabolism of ethanol varies considerably. Women have less of this enzyme than men and there are also marked ethnic variations. Most of the ethanol passes into the small intestine, where the majority of absorption takes place. The supply to the small intestine is governed by the rate at which the stomach empties, and this is an important variable affecting the delay between drinking and the consequent rise in blood alcohol concentrations. Food taken with alcohol, or immediately prior to it, will slow the rate of gastric emptying and encourage gastric metabolism, thus delaying peak blood alcohol concentrations and reducing their absolute levels.

Once in the small intestine alcohol is quickly absorbed and quickly distributed throughout the lean body mass. In women the increase in blood alcohol is generally higher for a given dose, since their proportion of body fat is on average slightly higher than that in men. In clinical studies the blood alcohol level peaks around 30–40 minutes after intake of a single dose (see Figure 5.4).

This is however unrepresentative of how alcohol is usually consumed in social drinking and other studies where intake is spread over a wider period show that blood alcohol can continue to rise for up to one hour after drinking.

In the body the main breakdown pathway for alcohol is by oxidation in the liver, catalysed by alcohol dehydrogenase together with NAD (the cofactor nicotinamide adenine dinucleotide) (Equation 5.2). The enzyme

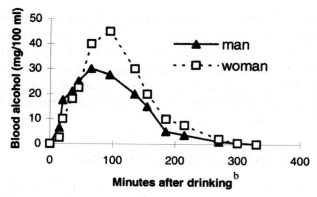

ᵃ J. E. Pieters, M. Wedel and G. Schaafsma, *Alcohol & Alcoholism*, 1990, **25(1)**, 17.
ᵇ Subjects consumed a single dose of 9.5 g alcohol.

Figure 5.4   *Blood alcohol levels (after Pietersᵃ)*

occurs in a number of isomeric forms of varying efficiency, which are under genetic control.

$$CH_3\text{-}CH_2\text{-}OH \overset{(a)}{\rightarrow} CH_3\text{-}CHO \overset{(b)}{\rightarrow} CH_3COO^- \rightarrow \text{Acetyl Coenzyme A}$$

$$NAD^+ \qquad\qquad NADH + H^+ \qquad\qquad\qquad (5.2)$$

(a) = alcohol dehydrogenase
(b) = aldehyde dehydrogenase

The second important pathway for ethanol metabolism is by the microsomal ethanol oxidising system (MEOS). This enzyme system uses reduced nicotinamide adenine dinucleotide phosphate (NADPH) and cytochrome P_{450} together with oxygen, and also produces acetaldehyde (Equation 5.3).

$$NADPH \qquad \boxed{Cyt\,P_{450}} \qquad NADP^+$$

$$CH_3\text{-}CH_2\text{-}OH \longrightarrow CH_3\text{-}CHO \qquad\qquad (5.3)$$

$$O_2 \qquad\qquad\qquad\qquad H_2O$$

This enzyme system is not normally present in cells but is induced in response to alcohol.

A third route for ethanol degradation is by the enzyme catalase, which is present in all cells (Equation 5.4). However, this is thought to be a minor route.

$$CH_3CH_2OH + H_2O_2 \xrightarrow{\text{catalase}} CH_3CHO + 2H_2O \qquad (5.4)$$

In each case the primary product of the reaction is acetaldehyde. This is a short-lived intermediate, being rapidly metabolised further to form acetate and thence acetyl coenzyme A (Equations 5.5 and 5.6).

$$CH_3CHO \xrightarrow{\text{acetaldehyde dehydrogenase}} CH_3COOH \qquad (5.5)$$

$$NAD^+ \qquad NADH + H^+$$

$$CH_3COO^- + \text{Coenzyme A} \xrightarrow{\text{thiokinase}} \text{Acetyl Coenzyme A} \qquad (5.6)$$

$$ATP \qquad AMP + pp_i$$

Acetyl coenzyme A is an essential intermediary which can shuttle 2-carbon units either into the Krebs cycle, where they are used for the production of energy or, if energy requirements are already being fulfilled, into fatty acid synthesis.

RISKS AND BENEFITS OF DRINKING ALCOHOL

Alcohol has been widely consumed in very many cultures for thousands of years and is valued for many qualities, but perhaps particularly for its social effects, in relaxing the mind and reducing inhibitions. However, it is also recognised that the excess consumption of alcohol, in whatever form, can have severe adverse effects on the health of an individual and frequently on their social and work-related circumstances as well. The majority of people who consume alcohol do so in moderate amounts and thus participate in a pleasurable activity which is a key part of social life in many communities. As discussed earlier in this chapter, there is increasing evidence that moderate consumption of alcoholic beverages is not only harmless, but may actually confer some health benefits. The key point, and one which remains the centre of much discussion and controversy, is where the dividing line comes between moderate and healthy, and immoderate and potentially harmful consumption. At what point are the protective effects against, for example, cardiovascular disease surpassed by the increased risks due to higher blood pressure and liver disease?

The balance between the protective and harmful health effects has been the subject of much discussion and there are several excellent reviews (for example, Doll,[8] Cooper[34]). Adverse health effects which are generally considered to be attributable to, or augmented by, high intake of alcohol include cirrhosis of the liver, cancers of the upper aero-digestive tract and haemorrhagic stroke.[35] There is also a weak association between alcohol intake and breast cancer, but as yet no plausible mechanism has been suggested.[36] Deaths and disabilities due to suicide and accidents also increase with alcohol intake. On the positive side, low to moderate consumption of alcohol is associated with fewer deaths from coronary heart disease and protective effects against some other vascular diseases and some infections, including stomach infections with *Helicobacter pylori*.[36] Protective effects have also been reported for gallstones[37] and kidney stones.[2,3] A number of studies suggest that moderate consumption of alcohol can also protect against osteoporosis in older women.

It must be recognised that there is unlikely to be a single rate of drinking, below which is completely safe for everyone and above which is equally harmful for everyone. For some people, in certain circumstances, the only completely safe rate might be very low or even abstinence. In this group would be, for example, individuals prescribed certain medicines where alcohol is inhibitory and people operating machinery or driving. Most health advisers recommend that pregnant women should drink less than is advised for women who are not pregnant. On the other hand, for those sections of the community who are at risk of coronary heart disease, the dividing line will be at higher intake levels.

Most countries issue official recommendations for alcohol consumption, usually in terms of drinks or units of alcohol per day or per week. These are consumption levels at which it is generally agreed there will very little risk of adverse effects. It does not follow that intakes slightly higher than those recommended are definitely harmful, but that there is an increased risk of harm. There is some variation between the recommendations in different countries, which only illustrates the lack of consensus amongst physicians.

In the UK one unit of alcohol is defined as 8 g of pure alcohol, which roughly equates to half a pint of normal strength (3.8%) beer, one average glass of wine or a public house measure of spirits. This unit, as defined, is smaller than in most other countries (for example, one unit in the USA is 12 g of alcohol), which must be taken into consideration when recommendations are compared. Table 5.10 shows recommendations for a number of countries. There has been a move, certainly in the UK, to give recommended intakes on a daily rather than a weekly basis, since there are indications that 'binge' drinking, is more harmful. There is a mounting

Table 5.10 *Definitions of units of alcohol in different countries*

Country	Definition of a unit (g alcohol)	Official recommendations
Australia (set 1992)	1 drink = 8–10	Men, <4 drinks/day Women <2 drinks/day
Austria	1 unit = 6.3	
Canada	1 drink = 13.6	Men, 2 drinks/day Women, 0.7 drinks/day
Denmark	1 drink = 12	Men, <21 units/week Women, <14 units/week
Japan	1 unit = 19.75	
New Zealand		Men, 3–4 drinks/day Women 2–3 drinks/day
Sweden		Men & Women, <50 g alcohol/week
UK (set 1995)	1 unit = 8	Men, <4 units/day Women <3 units/day
USA	1 drink = 12	Men, 2 drinks/day Women, 1 drink/day

realisation that the Mediterranean habit of regular daily consumption of moderate amounts of alcohol, usually with meals, may be better for us than the old British tradition of getting drunk on Friday or Saturday night and not touching alcohol for the rest of the week. Hopefully this is a tradition now in terminal decline!

POTENTIAL FOR FUTURE DEVELOPMENT

As is evident from the information in this chapter, evidence is now accumulating to show that the raw materials from which beer is made contain a wide range of trace compounds which could be beneficial to health, over and above the major nutrients, which are well recognised. These compounds are generally present in beer in quite small quantities, but the possibility exists that in the future specific beer-based products could be designed to fulfil specific health needs. In other words, beer could provide a functional food. One example might be a beer which is enriched with antioxidant vitamins, or with hop flavonoids to protect against osteoporosis or cancer. Or even a fibre-rich beer, which could provide a palatable and unusual alternative to existing high-fibre compounds!

SUMMARY

Beer is a well balanced package containing moderate amounts of alcohol as well as carbohydrate, protein, fibre and useful amounts of B vitamins and essential minerals. It is low in sugar, salt and fat. In addition, it contains trace ingredients such as anti-oxidants which may help to protect against certain degenerative diseases. Beer is available in a wide range of styles and flavours. When consumed in sensible amounts it can contribute to a healthy diet.

REFERENCES

1 A. Piendl, *Brauwelt*, 1990, **130(11)**, 370.
2 J. Shuster, B. Finlayson, R.L. Scheaffer, R. Sierakowski, J. Zoltek and S. Dzegede, *J. Chronic Dis.*, 1985, **38**, 907.
3 G.C. Curhan, W.C. Willett, F.E. Speizer and M.J. Stampfer, *Annals Internal Med.*, 1998, **128(7)**, 535.
4 M.J. Thun, R. Peto, A.D. Lopez, J.H. Monaco, S.J. Henley, C.W. Heath and R. Doll, *New Eng. J. Med.*, 1997, **337(24)**, 1705.
5 R. Doll, R. Peto, E. Hall, K. Wheatley and R. Gray, *BMJ*, 1994, **309**, 11.
6 J-M. Yuan, R.K. Ross, Y-T. Gao, B.E. Henderson and M.C. Yu, *Br. Med. J.*, 1997, **314(7073)**, 18.
7 C. D'Arcy, J. Holman, D.R. English, E. Milne and M.G. Winter, *Med. J. Aus.*, 1996, **164(3)**, 141.
8 R. Doll, *Drug and Alcohol Review*, 1998, **17**, 353.
9 J. Chick, *Alcohol and Alcoholism*, 1998, **33(6)**, 576.
10 B.A. Clevidence, M.E. Reichman, J.T. Judd, R.A. Muesing, A. Schatzkin, E.J. Schaefer, Z. Li, J. Jenner, C.C. Brown, M. Sunkin, W.S. Campbell and P.R. Taylor, *Arteriosclerosis, thrombosis and vascular biology*, 1995, **15(2)**, 179.
11 D.F. Jansen, S. Nedeljkovic, E.J.M. Feskens, M.C. Ostojic, M.Z. Grujic, B.P.M. Bloemberg and D. Kromhout, *Arteriosclerosis, thrombosis and vascular biology*, 1995, **15(11)**, 1793.
12 D.R. Parker, J.B. McPhillips, C.A. Derby, K.M. Gans, T.M. Lasater and R.A. Carleton, *Am. J. Public Health*, 1996, **86(7)**, 1022.
13 M. Stefanick, C. Legault, R.P. Tracy, G. Howard, C.M. Kessler, D.L. Lucas and T.L. Bush, *Arteriosclerosis, thrombosis and vascular biology*, 1995, **15(12)**, 2085.
14 S. Renaud, A.D. Beswick, A.M. Fehily, D.S. Sharp and P.C. Elwood, *Am. J. Clinical Nutr.*, 1992, **55**, 1012.
15 J-C. Hsu and D.A. Heatherbell, *Am. J. Enol. Vitic.*, 1987, **38(1)**, 6.

16 E. Pueyo, M. Dizy and M. CarmenPolo, *Am. J. Enol. Vitic.*, 1993, **44(3)**, 255.

17 J.S. Hough, D.E. Briggs, R. Stevens and T.W. Young, in *Malting and Brewing Science*, Chapman and Hall, London, 1982, Volume 2.

18 McCance and Widdowson's *The Composition of Foods*, First supplement, by A.A. Paul, D.A.T. Southgate and J. Russell, HMSO, 1980.

19 R.K. Newman, C.W. Newman and H.Graham, *Cereals Foods World*, 1989, **34(10)**, 883.

20 National Food Survey 1996. Annual report on food expenditure, consumption and nutrient intakes. MAFF, 1997.

21 J.P. Bellia, J.D. Birchall and N.R. Roberts, *Annals Clin. Lab. Sci.*, 1996, **26(3)**, 227.

22 C.N. Martyn, *Lancet*, 1990, Aug 18th, 430.

23 N.B. Roberts, A. Clough, J.P. Bellia and J.Y. Kim, *J. Inorganic Biochem.*, 1998, **69**, 171.

24 M.S. van der Gaag, J.B Ubbink, P. Sillanaukee, S. Nikkari and H.F.J. Hendriks, *Lancet*, 2000, **355(9214)**, 1522.

25 M. Law and N. Wald, *BMJ*, 1999, **518**, 1471.

26 E.B. Rimm, A. Klatsky, D. Grobbee and M.J. Stampfer, *Br. Med. J.*, 1996, **312(7033)**, 731.

27 A. Ghiselli, F. Natella, A. Guidi, L. Montanari, P. Fantozzi and C. Scaccini. *J. Nutr. Biochem.*, 2000, **11(2)**, 76.

28 L. Bourne, G. Paganga, D. Baxter, P. Hughes and C.A. Rice-Evans, *Free Radical Res.*, 2000, **32**, 273.

29 C.L. Miranda, J.F. Stevens, A. Sharps, M.C. Henderson and R.J. Rodriguez, *J. Am. Soc. Brew. Chemists*, 1998, **56(4)**, 136.

30 M.C. Henderson, C.L. Miranda, J.F. Stevens, M.L. Deinzer and D.R. Buhler, *Xenobiotica*, 2000, **30(3)**, 235.

31 D.C. Knight and J.A. Eden, *Obstetrics and Gynaecology*, 1996, **87**, 897.

32 S.R. Milligan, J.C. Kalita, A. Heyerick, H. Rong, L. De Cooman and D. De Keukeleire, *J. Clin. Endocrin. Metabolism*, 1999, **83**, 2249.

33 W.J. Simpson, *J. Inst. Brewing*, 1993, **99(5)**, 405.

34 M.Gurr, in *Alcohol, Health issues related to alcohol consumption*, ILSI Europe Concise Monograph series, Brussels, 1996.

35 T.J. Cooper, *Proc. 23rd Conv. Inst. Brewing (Asia Pacific Section)*, *Sydney*, 1994, 32.

36 *Sensible Drinking*. Report of an Inter-departmental Working Group, Department of Health, December 1995.

37 H. Brenner, D. Rothenbacher, G. Bode and G. Alder, *Br. Med. J.*, 1997, **315**, 1489.

38 F-X. Caroli-Bosc and others, *Digestive Diseases and Sciences*, 1998, **43(9)**, 2131.

Chapter 6

Assuring the Safety of Beer

RISKS TO FOOD SAFETY

For any foodstuff, the potential risks can be grouped into a relatively small number of categories:

- Natural components of the raw materials which are themselves inherently toxic
- Environmental contaminants associated with the raw materials
- Microbiological infections
- Contaminants derived from transport, distribution, storage or packaging
- Contaminants derived from the processing ·
- Contaminants associated with additives, processing aids or other materials used in processing which may come into contact with the food
- Deliberate contamination of a foodstuff with a harmful material
- Components such as allergens which, although harmless to the majority of the population, may still pose substantial risk to a small minority

Beer is no exception to these risks. However, some are less of a concern. For example, one of the most serious risks with many foods is the potential for contamination by food poisoning bacteria, such as *Clostridia* or *Listeria*. Beer is, as has been mentioned in earlier chapters, inherently a very safe food microbiologically speaking. This is partly because of the boiling stage, which essentially kills any microbiological contaminants arising from the raw materials, and also because of the anti-bacterial effects of the alcohol, the low pH, the carbon dioxide and the hop acids. This does not mean that infections in beer are impossible, just that they are unlikely to be dangerous.

Different strategies must be applied to the categories of hazards described above in order to eliminate or minimise them. For example, while a contaminant derived from the transport or distribution chain could, theoretically at least, be completely eradicated, it may be biologically impossible to eliminate a toxicant that is an integral component of a certain foodstuff. In the latter case controls must be devised and imposed in order that concentrations of the toxic material are maintained at acceptable levels. It also has to be realised that nothing in life can be guaranteed as 100% safe. A number of materials, including the most basic requirements of oxygen, food and water, are essential for life at one concentration, but can be highly dangerous at another. The old adage 'the dose makes the poison' still holds.

There is also the concept of acceptable – or unacceptable – risk. We know that driving very fast is dangerous and likely to result in accidents. Yet a significant number of us persist in driving above legal or recommended limits and cars which can reach ever faster speeds are constantly being produced. Likewise we know that chocolate and sweets may be bad for our teeth and can result in our putting on too much weight. We recognise those risks and moderate our intake accordingly, depending upon the seriousness with which we view them. In the case of an alcoholic beverage such as beer, there is a recognised health risk associated with excess consumption. The question is sometimes asked as to why the presence of low levels of chronic toxicants should be a concern, since the amount of the beverage which would need to be consumed to cause health problems from such low level contaminants would undoubtedly cause problems from the alcohol alone. The answer to this is that the alcohol is a recognised risk, the quantities are defined and declared and consumers are free to make their own choices. Contaminants are, by definition, not intended to be in the product and therefore the consumer is unaware of them, and cannot exercise his own judgement or free will. Alcoholic beverages are subject to the same food safety legislation as any other food. The rule for alcoholic beverages, both legally and morally, is exactly the same as for any other food – there should be no avoidable harmful materials present and those that are unavoidable (such as natural toxicants) must be reduced to the lowest possible levels.

HACCP

The approach used by most brewers to maximise food safety and to minimise risks to consumers is the same as that used in a large sector of the food industry and is known by its initials, HACCP.

The term HACCP means **H**azard **A**nalysis by **C**ritical **C**ontrol **P**oints and comes originally from the systems devised to ensure that the food used by American astronauts was absolutely safe from the microbiological point of view. Nowadays it is in widespread use in the food industry, to ensure chemical as well as microbiological safety and is often extended to cover quality parameters as well. A HACCP system can be described as follows:

- The whole of the process must be considered, from the raw materials used to the final products produced. A simple flow diagram (see Figure 6.1) is useful here
- The potential hazards associated with all the materials used and with each stage of production must be listed. The likelihood of each hazard actually occurring will be taken into account in order to identify the real risks to products
- Next, the process stage at which each of these risks can most effectively be controlled must be identified (the Critical Control Points) and a suitable control system must be put in place to ensure that any undesirable effects are eliminated or controlled to acceptable levels (see Figure 6.2)
- There must be some way of checking that the controls are working and the acceptable limits must be specified
- If these limits are exceeded then corrective actions will need to be taken. These should be specified. There should also be some way of checking that these corrective actions have been followed and that they have achieved the desired results

Records should be kept so that it is possible to take any batch of beer and confirm that the raw materials were to the defined quality specifications and that the process was working within the defined limits.

Example – Wort Boiling. Before fermentation, the wort is boiled with hops or hop products. This process is important for several reasons:

1. It converts the hop acids into the isomerised (bitter) form,
2. It coagulates unwanted proteins,
3. It volatilises unwanted flavour compounds,
4. It sterilises the wort.

It is therefore crucial that temperatures close to 100 °C are achieved for the required time. The exact temperature of boiling is not always 100 °C but will vary depending upon the atmospheric pressure. The boil temperature should therefore be measured and recorded, and if this is outside the specified limits (perhaps ± 2 °C) the boil time will need to be extended or the wort kettle sealed to increase the internal pressure and thus raise the boiling point of the wort.

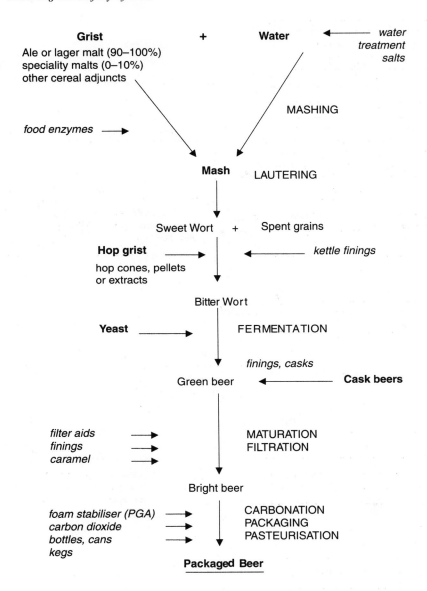

Figure 6.1 *Flow diagram for the brewing process*

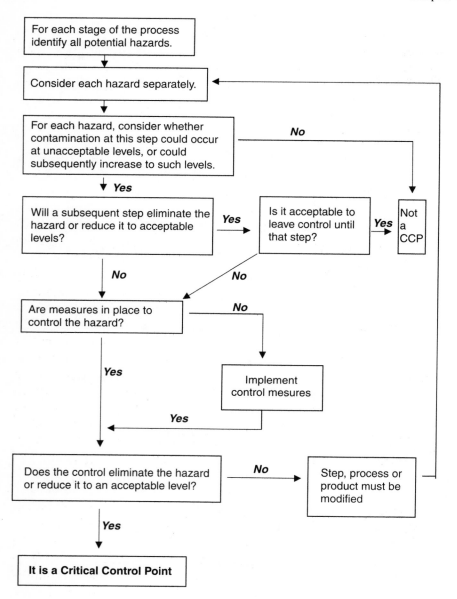

Figure 6.2   *Decision tree for Critical Control Points in brewing*

Raw Materials

The main raw materials for brewing are malted barley, hops (or hop products such as pellets or extracts), yeast and water. Other cereals, either malted or unmalted, in the form of flours, grits, or whole grain, may also be used for some beers to impart specific characteristics. Cereal syrups can also be used as an additional source of fermentable sugars.

Cereals and Malt

The chief risks for these materials are chemical contaminants, fungal infections and insect infestations. Legal limits are set for contaminants such as pesticide residues and heavy metals (such as lead) in agricultural crops and a brewer will insist that malt supplied to him is prepared from cereals that comply with any relevant legislation. Storage conditions are also important: malt especially is very vulnerable since it can readily absorb taints. Maximum limits will also be set for volatile nitrosamines. The detection of these carcinogenic contaminants in beer and whisky (as well as in certain other foods such as cured meats) during the early 1970s caused much concern and disruption in the malting and brewing industries, until the source of the contamination was identified. After intensive investigations it was found that NDMA (*N*-nitrosodimethylamine) can be formed during malt kilning by reaction between oxides of nitrogen (NO_x) in kiln gases and amines in the grain (see Figure 6.3). In a modern malting plant the formation of nitrosamines is avoided by the use of indirectly fired kilns, in which kiln gases do not come into contact with the grain bed. Levels of NDMA are now very low, but both malts and beers continue to be routinely monitored to ensure that nitrosamine content is within specified limits.

Fungi can infect grain both in the field and during storage. Different species are involved in each case, adapted to the moisture available, which ranges from 40–50% in the ripening kernel, to below 18% in the ripe grain. Such fungal infections must be avoided where possible, both because of the damage they can cause to grain viability and beer quality, but also because they can be associated with the formation of toxic secondary metabolites, known as mycotoxins. The most toxic of these, the aflatoxins, are largely confined to crops grown in tropical climates, because the fungi which produce them require higher temperatures and humidity than is normally encountered in Europe. Legal controls have been introduced by the EU in order to guard against possible contamination of imported produce, especially peanuts and maize. These impose a limit of $4\,\mu g\,kg^{-1}$ for total aflatoxins in cereals, and a limit of $2\,\mu g\,kg^{-1}$

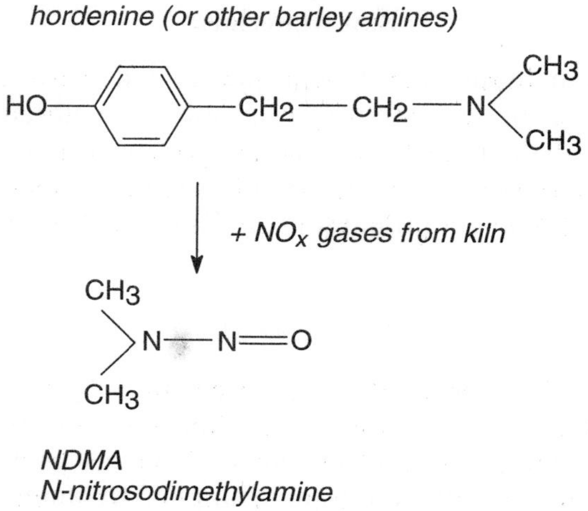

hordenine (or other barley amines)

+ NO$_x$ gases from kiln

NDMA
N-nitrosodimethylamine

Figure 6.3   *Formation of volatile nitrosamines during directly fired kilning*

for the most toxic one, B$_1$.

The most common toxin-producing fungal species which occur in the field in Europe and North America are the *Fusaria*. These are widespread in cereals, especially maize and wheat as well as barley. Infection levels are generally low, although floods in the grain producing areas of the American mid-west in the late 1990s resulted in relatively high levels in some crops. *Fusarium* moulds can produce a number of mycotoxins. Chemically these are known as trichothecenes, and include such compounds as deoxynivalenol, nivalenol and T-2 toxin. These are substantially less toxic than the aflatoxins, and at the time of writing there are no legal limits in the EU, although 'action levels' of between 500 and 750 μg kg^{-1} for cereals and cereal products are under discussion. In the malting industry, the appearance and odour of the cereal grains is checked on a grain sample before purchase and again at intake. Bulks showing signs of mould infection will be rejected. As a consequence, levels of *Fusarium* toxins in malting quality barley are generally well below the proposed 'action levels' and below those in feed quality grain.

Other moulds, such as *Penicillium* and *Aspergillus* species can flourish at the lower moisture contents of ripe grain (between 16 and 20% moisture). Under suitable conditions of temperature and moisture these species can produce another mycotoxin, ochratoxin A. Ochratoxin A is thought to be carcinogenic, and legal limits for grain are currently being introduced by the EU (3–5 μg kg^{-1}). Barley for malting is routinely dried to below 14% moisture (often to below 12% for long term storage) in order to conserve the viability, since this is of course essential for germination. These moisture levels are too low to support mould growth, especially since the grain is also usually cooled for long term storage. Malt itself is kilned to a moisture content of not more than 6%, which, provided that it is properly stored in dry conditions, is also too dry to permit fungal growth. Levels of ochratoxin A are routinely monitored in malting barley and in malt but are always well below the proposed legal limits.

Insect infestations are always a potential problem where grain is stored for any length of time. Attention to cleanliness and hygiene in grain stores is essential. Drying and cooling are also important, since many insects cannot breed below certain moistures and temperatures. Unmalted cereals, which may be in store for up to a year between harvests, may be treated with approved insecticides. This is not allowed for malt. Although partially protected from insect attack by its low moisture, malt is also manufactured and traded throughout the year as required, to avoid the need for long-term storage.

Any HACCP plan will, therefore, pay particular attention to checking storage conditions for all raw materials.

One group of environmental contaminants that increasingly causes problems is the PCBs (polychlorinated biphenyls) and the structurally related dioxins. These organic contaminants are derived almost entirely from man-made sources. One particularly important source is the incineration of waste materials, which, if not carried out at sufficiently high temperatures, can generate both PCBs and dioxins, both of which are extremely toxic chemicals and can contaminate pastures and crops in the surrounding areas. Contamination of animal feed in Belgium, and consequently of a number of animal byproducts used in food manufacturing, caused a major international incident in Europe in 1999. PCBs and dioxins, like many of these organic contaminants, are lipid soluble and therefore tend to concentrate in high fat foods, such as milk, oils and some cuts of meat. Beer is particularly protected in this respect, since barley is relatively low in lipid (about 3% of the barley grain is lipid) and beer, as has already been described in Chapter 5, contains negligible amounts of lipid.

Hops and Hop Products

As with cereals, the main risks are from residues of pesticides and heavy metals. Limits may also be set for nitrates, which tend to accumulate in the leafy parts of any plant, and can be very high in hop cones. Processing, for example to produce hop extracts in which the hop acids are concentrated, generally reduces the concentration of such contaminants significantly.

Water

Over 90% of beer is water. It is not surprising, therefore, that the quality
of that water is of paramount importance for the quality of the final
product. Water used for brewing, whether from the brewery's own wells
or from the mains, must be of drinking water quality with regards to any
harmful chemical and microbiological contaminants. This means that it
must comply with any current legislation. Within the EU, drinking water
legislation sets limits for chemical and microbiological contaminants and
specifies the frequency of testing for each parameter (Table 6.1). Heavy
metals, such as lead and cadmium are a major concern in water for food
manufacturing. Here, however, beer is protected, since heavy metals bind
to the spent grains, the trub and the waste yeast and are removed from the
process stream. Indeed, some systems which have been designed for
cleaning up contaminated water supplies use columns of yeast to absorb
heavy metals. Thus beer normally contains lower levels of these contaminants than the water used in its manufacture. With organic contaminants, however, legislation does not always provide enough quality
assurance for the brewer, since it allows some test results to be averaged
over a period of time. Thus pollutants which originate from single point
sources – chlorinated solvents from dry-cleaning processes are a typical
example of this – can exceed safety or quality limits for only short periods
of time, but are sufficient to contaminate one or more brews. In order to
avoid this many breweries, particularly those situated in vulnerable
urban areas, have installed sophisticated water treatment plants, for
example ion exchange columns or reverse osmosis systems. Where these
are installed, they must be included within the HACCP plan. Thus there
should be clear procedures for maintenance and cleaning of any filters
and regular monitoring to ensure that they are working within specified
parameters.

Yeast

Most breweries reuse their yeast several times; consequently the chief
safety issue is to avoid microbiological contamination, either with bacteria or with wild yeasts. A HACCP plan will therefore specify how yeast
is to be handled and stored after it is cropped at the end of fermentation.
In large breweries, there may be a set number of generations allowed
before a fresh culture is introduced. In many smaller breweries, however,
the yeast has been re-pitched from one fermentation into the next for
many generations – that is, generations of Head Brewers, let alone yeast!
These breweries often rely upon a mixture of related yeast strains, which

Table 6.1 *Safety and quality requirements for water used in brewing*

Parameter	Units	Maximum concentration
Toxic Chemicals		
Arsenic	$\mu g\,l^{-1}$	50
Cadmium	$\mu g\,l^{-1}$	5
Cyanide	$\mu g\,l^{-1}$	50
Chromium	$\mu g\,l^{-1}$	50
Mercury	$\mu g\,l^{-1}$	1
Nickel	$\mu g\,l^{-1}$	50
Lead	$\mu g\,l^{-1}$	50*
Antimony	$\mu g\,l^{-1}$	10
Selenium	$\mu g\,l^{-1}$	10
Pesticides and related products		
individual substances	$\mu g\,l^{-1}$	0.1
total substances	$\mu g\,l^{-1}$	0.5
Polycyclic aromatic hydrocarbons	$\mu g\,l^{-1}$	0.2
Trihalomethanes	$\mu g\,l^{-1}$	100
Quality parameters		
Sulfate	$mg\,SO_4\,l^{-1}$	250
Magnesium	$mg\,Mg\,l^{-1}$	50
Sodium	$mg\,Na\,l^{-1}$	150
Potassium	$mg\,K\,l^{-1}$	12
Nitrate	$mg\,NO_3\,l^{-1}$	50
Nitrite	$mg\,NO_4\,l^{-1}$	0.1
Ammonium	$mg\,NH_4\,l^{-1}$	0.5
Hydrocarbons	$\mu g\,l^{-1}$	10
Phenols	$\mu g\,C_6H_5OH\,l^{-1}$	0.5
Surfactants	$\mu g\,l^{-1}$	200
Aluminium	$\mu g\,Al\,l^{-1}$	200
Iron	$\mu g\,Fe\,l^{-1}$	200
Manganese	$\mu g\,Mn\,l^{-1}$	50
Copper	$\mu g\,Cu\,l^{-1}$	3000
Zinc	$\mu g\,Zn\,l^{-1}$	5000
Phosphorus	$\mu g\,P\,l^{-1}$	2200
Fluoride	$\mu g\,F\,l^{-1}$	1500
Silver	$\mu g\,Ag\,l^{-1}$	10
Microbiological parameters		
Total coliforms	number/100 ml	0
Faecal coliforms	number/100 ml	0
Faecal *Streptococci*	number/100 ml	0
Sulfite-reducing *Clostridia*	number/20 ml	$\leqslant 1$

Source: The Water Supply (Water Quality) Regulations 1989; UK Statutory Instrument 1989 No. 1147.
*This will be reduced to 25 and then to $10\,\mu g\,l^{-1}$ in draft amending legislation.

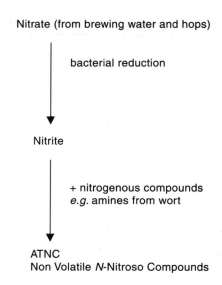

Nitrate (from brewing water and hops)

bacterial reduction

Nitrite

+ nitrogenous compounds
e.g. amines from wort

ATNC
Non Volatile *N*-Nitroso Compounds

Figure 6.4   *Formation of non-volatile nitrosamines*

has evolved over decades to suit the process conditions and product quality required, adding to the complication of trying to use a pure culture. Provided that this yeast is properly managed and stored so that it is fully viable, it will easily out-grow any competing spoilage organisms, and can be used almost *ad infinitum* without any detrimental effects on beer quality. Fortunately, as has already been said, the combination of hop bitter acids, low pH, carbon dioxide concentrations and anaerobic conditions associated with fermenting beer mean that few, if any, pathogenic bacteria can grow. The main concern is with spoilage bacteria. These are bacteria that have adapted to the conditions in beer and, although not harmful to man, can produce hazes or off-flavours in the beer. Some spoilage bacteria, such as *Obesumbacteria proteus* can also be a safety concern indirectly, since they are able to reduce nitrates to nitrite, which in turn can form nitrosamines (Figure 6.4). Very similar reactions occur in the human gut. Unlike NDMA, these nitrosamines are non-volatile and form a group of compounds which have not yet been completely characterised. They are therefore usually referred to as ATNC, or Apparent Total *N*-nitroso Compounds. To avoid any ATNC formation, some brewers have introduced regular washing of the yeast with dilute food grade acid in order to kill any bacteria. It is also becoming increasingly common to limit the number of generations for which the yeast is reused.

A good HACCP plan for a small brewery will include regular examination of the yeast by light microscope to look for wild yeasts and contaminating bacteria. It will also use tasting and quality analysis data from the finished beer to monitor for the presence of spoilage bacteria, since taints produced by such organisms can generally be detected by trained tasters at extremely low concentrations.

Processing

The HACCP plan must also consider any hazards associated with the processing itself. This will include:

- The brewing plant, which should be of a hygienic design, with no 'dead legs' in the pipework which cannot be adequately cleaned.
- Materials of construction, which must be resistant to corrosion and soiling, able to be completely cleansed by the type of cleansers approved for food contact materials, and must not leach materials which could be harmful or could impart taints to the beer.
- Cleansers and sterilising materials: these must be approved for food contact materials and must not leave harmful residues or taints, or otherwise damage the beer (for example by affecting foam).
- Frequency and efficiency of cleaning and rinsing cycles.
- Use of any additives or processing aids: these must be approved for the use envisaged, manufactured to the correct specifications and added at the correct dose levels.
- Critical parameters, such as times, temperatures and extent of mixing.

The HACCP system must ensure that suitable controls are in place to keep the process working in the way for which it is intended. It must also ensure that, should any deviations arise, they are detected and corrected. Again, adequate records must be kept so that the exact processing conditions for any one batch can be traced subsequently.

Microbiological Safety

As mentioned earlier, the inability of pathogenic bacteria to grow in beer means that brewing is not a high risk industry in terms of microbiological safety. The quality of the product can however be compromised by infections with spoilage bacteria. Although the boiling stage after mashing effectively sterilises the wort prior to fermentation, bacterial activity pre-boil is undesirable since some of the products of such activity – such as nitrite or toxins – are potentially harmful. For example, any nitrite can react with amines in the wort, as already described for spoilage bacteria at the start of fermentation, producing non-volatile nitrosamines, ATNCs. The HACCP plan must therefore address the hygiene of the mashing plant and ensure that the specified mash and boil temperatures are achieved for the required times. Worts which are held for any period of time before boiling, or which are to be recycled into the mashing vessel must be maintained at a high enough temperature to prevent bacterial activity.

Standards of cleanliness are even higher for processing downstream of the boil stage. Beer that is destined for packaging in bottles, cans or kegs has the added safeguard of pasteurisation, but, especially in the UK, a substantial volume of beer is still sold unpasteurised in casks. The HACCP plan will also need to take into account any products or processes which are particularly vulnerable, such as low alcohol brands and refillable bottles. The operation of pasteurisers is obviously critical, and controls need to be in place to check that the correct operating temperatures are always achieved and that dwell times are sufficient. The HACCP plan will specify the frequency and type of microbiological tests carried out and when, for example, rapid tests such as ATP bioluminescence (which can give a measure of bacterial contamination within a few minutes rather than several days as required for conventional tests) should be utilised.

Packaging

In modern breweries packaging is a very high speed operation – 2000 bottles a minute is not uncommon – and consequently vulnerable. The HACCP plan must encompass:

- Quality of packaging materials – they should be suitable for food use and should not leach harmful or tainting chemicals into the product
- Storage of empty packages, in order to avoid pickup of dirt or taints
- Adequate cleaning of refillable packages
- Possibility of foreign objects (insects for example) getting into the package during filling
- Possibility of glass in package: this could originate from faulty bottles or from filler-heads and sight glasses
- Malicious tampering
- Cleanliness of fillers
- Possibility of contamination, for example with lubricants (from conveyors *etc.*), coolants or cleansers
- Efficient operation of pasteurisers
- Sterile filling

Packaging equipment nowadays is usually fitted with a large number of built-in checking systems and safeguards, for example to check for the presence of foreign objects in the package. The HACCP system must ensure that there is adequate proof that these are working efficiently and that they can detect all inclusions. For example, it is common practice to utilise a number of 'test' packages with inclusions at different heights in the bottle or can to check that detection systems are working.

Deliberate Tampering

There are a number of measures that can be taken to reduce the possibility of deliberate introduction of harmful materials into food. Most of these measures are common to foods in general. They include security at manufacturing and storage sites, covered and/or sealed vessels, and the use of 'tamperproof' packaging. The nature of most beer packaging, that is, cans, bottles or kegs, means that any interference with the finished package is likely to be evident. It is however impossible to totally prevent access to process streams during manufacture. Here the main safeguard has to be the complete traceability of the product, by lot marking, packaging codes and the accompanying documentation, so that any suspect batches can be rapidly identified and withdrawn, wherever they are in the distribution system. Fortunately such incidents are rare, although when they do occur, they tend to attract a great deal of publicity.

Allergens

Two potential allergens must be considered in relation to beer. One is gluten, which is the name given to the complex of gliadin proteins and carbohydrate found in wheat. Structurally similar proteins are found in the closely related cereals barley, oats and rye. The related protein in barley is known as hordein, from the Latin name for barley, *Hordeum*. These proteins can irritate the cells lining the stomach of certain people, causing coeliac disease, and coeliac sufferers must avoid consuming any foods which contain wheat derivatives or any other of the offending cereals. Barley and malt are obviously harmful to coeliac sufferers. There is some debate as to whether beer is also harmful, since a substantial proportion of barley protein is left behind in the spent grains. Also, much of what is extracted into the wort is later removed as trub or is filtered out before packaging, thus does not persist into the beer. Chemical analysis of beer or other processed food for gluten is subject to a number of limitations and the presence of coeliac-positive proteins in beer has not been proved conclusively, although some barley-derived peptides are certainly

Table 6.2 *Levels of sulfites allowed in some foods*

Food	Maximum level of sulfite, expressed as SO_2 $mg\,l^{-1}$ or $mg\,kg^{-1}$
Beer	20
Cask conditioned beer	50
Wine	160–400 (depending upon type of wine)
Cider	200
Fruit juices (catering use)	50
Breakfast sausages	450

Sources: Council Regulation (EEC) No. 822/87 of 16 March 1987 on the common organization of the market in wine.
European Parliament and Council Directive 95/2/EC on food additives other than colours and sweeteners.
The Miscellaneous Food Additives Regulations 1995.

present. Nevertheless, coeliac sufferers are advised to sample beer with care, since sensitivities can vary widely between individuals.

The other allergenic reaction (more properly described as a food intolerance) which may be associated with beer is a reaction to sulfur dioxide. This is an approved additive which has been used as a preservative in a wide range of foodstuffs for many years and, at the levels used, poses no health hazards to the great majority of people. However, a small number of individuals are hypersensitive to sulphite and these people may suffer severe asthmatic reactions, which may even be fatal, even at low levels of exposure. Sulfur dioxide can be added, usually in the form of sodium or potassium metabisulfite, to a wide range of foodstuffs, including alcoholic beverages, as an antioxidant and preservative. In beer the main function is to protect the beer from formation of stale oxidised flavours (see Chapter 4). The levels of sulphite are controlled by EU law (see Table 6.2), with the limit for packaged beers being $20\,mg\,kg^{-1}$ ($50\,mg\,kg^{-1}$ are allowed in cask-conditioned beers), which is relatively low compared with some other foods.

SUMMARY

Beer is a low risk foodstuff. This can be attributed to:

- Its physical and chemical properties, which render it inherently more resistant to microbiological attack (alcohol content, low pH, carbon dioxide content and low oxygen tension)
- Use of wholesome raw materials (sprouted cereals and hops)
- A boiling stage which kills any microbiological contaminants associated with the raw materials

- Hygienic processing and widespread use of independently accredited quality systems
- The solid/liquid separation stages, which eliminate or reduce many relatively insoluble contaminants
- The ability of yeast to bind ionic contaminants such as heavy metals

FURTHER READING

1 E.D. Baxter, *New Food*, 1999, **2(4)**, 27.
2 G. Jackson, 'Symposium on quality issues and HACCP', *Eur. Brew. Conv. Monog. 26*, Stockholm, 1997, 50.
3 D.E. Long, *J. Inst. Brewing*, 1999, **105(2)**, 79.
4 A. Mundy, *Brewer*, 1997, **83(997)**, 517.
5 A. Mundy, 'Symposium on quality issues and HACCP', *Eur. Brew. Conv. Monog. 26*, Stockholm, 1997, 141.

Subject Index